AF786063

Schriften zu Leben und Gesellschaft - Series on Life and Society

Volume 3

Series Editors

Björn Schmitz-Luhn, Center for Life Ethics, Rheinische Friedrich-Wilhelms-Universit, Bonn, Germany

Christiane Woopen, Center for Life Ethics, Rheinische Friedrich-Wilhelms-Universit, Bonn, Germany

Über die Schriftenreihe Wissenschaft und Gesellschaft stehen durch zahlreiche Veränderungsprozesse und Krisen vor immer wieder neuen Herausforderungen. Technologisierung, Ökonomisierung, Ökologisierung und Globalisierung setzen das individuelle und gesellschaftliche Leben unter Gestaltungsdruck. Eine Vielfalt wissenschaftlicher Disziplinen und Akteure aus allen gesellschaftlichen Bereichen sind gefordert, die transformativen Prozesse der Gegenwart gemeinsam, kreativ und kritisch reflektierend zu gestalten. Die Schriften zu Leben und Gesellschaft stellen das Leben in den Mittelpunkt—das Leben nicht ausschließlich in einem biologisch-naturwissenschaftlichen Verständnis, sondern in allen Facetten des sich entfaltenden und entwickelnden Lebens. Es geht um das menschliche und das nichtmenschliche Leben, um das individuelle und das gesellschaftliche Leben, jenseits disziplinärer Grenzen. In ausgewählten Studien und Analysen präsentieren sie fachübergreifende und translationale Forschung, Debatten und Konzepte zur partizipativen Gestaltung unserer Zukünfte im globalen Kontext.

Stefan Partelow
Editor

Ethics and the Commons

Building Ethical Awareness in Research and Practice

Springer

Editor
Stefan Partelow
Center for Life Ethics
Bonn, Germany

Faculty of Agricultural, Nutritional
and Engineering Sciences (AEI)
University of Bonn
Bonn, Germany

ISSN 3005-0901 ISSN 3005-091X (electronic)
Schriften zu Leben und Gesellschaft - Series on Life and Society
ISBN 978-3-032-24245-7 ISBN 978-3-032-24246-4 (eBook)
https://doi.org/10.1007/978-3-032-24246-4

This work was supported by Center for Life Ethics (101164978).

© The Editor(s) (if applicable) and The Author(s) 2026. This book is an open access publication.

Open Access This book is licensed under the terms of the Creative Commons Attribution-NonCommercial-NoDerivatives 4.0 International License (http://creativecommons.org/licenses/by-nc-nd/4.0/), which permits any noncommercial use, sharing, distribution and reproduction in any medium or format, as long as you give appropriate credit to the original author(s) and the source, provide a link to the Creative Commons license and indicate if you modified the licensed material. You do not have permission under this license to share adapted material derived from this book or parts of it.
The images or other third party material in this book are included in the book's Creative Commons license, unless indicated otherwise in a credit line to the material. If material is not included in the book's Creative Commons license and your intended use is not permitted by statutory regulation or exceeds the permitted use, you will need to obtain permission directly from the copyright holder.
This work is subject to copyright. All commercial rights are reserved by the author(s), whether the whole or part of the material is concerned, specifically the rights of translation, reprinting, reuse of illustrations, recitation, broadcasting, reproduction on microfilms or in any other physical way, and transmission or information storage and retrieval, electronic adaptation, computer software, or by similar or dissimilar methodology now known or hereafter developed. Regarding these commercial rights a non-exclusive license has been granted to the publisher.
The use of general descriptive names, registered names, trademarks, service marks, etc. in this publication does not imply, even in the absence of a specific statement, that such names are exempt from the relevant protective laws and regulations and therefore free for general use.
The publisher, the authors and the editors are safe to assume that the advice and information in this book are believed to be true and accurate at the date of publication. Neither the publisher nor the authors or the editors give a warranty, expressed or implied, with respect to the material contained herein or for any errors or omissions that may have been made. The publisher remains neutral with regard to jurisdictional claims in published maps and institutional affiliations.

This Springer imprint is published by the registered company Springer Nature Switzerland AG
The registered company address is: Gewerbestrasse 11, 6330 Cham, Switzerland

If disposing of this product, please recycle the paper.

Foreword

As the authors of Ethics and the Commons state in their summary chapter: "Ethics guide all intentional human actions, including commons governance and commoning."

This sentence resonated with me deeply. I began my forest research in 1994 as an ecologist seeking to understand the design principles that facilitated community forest management. It was much later, in 2007, that I engaged with the commons as a mother and a resident—working with activist groups engaged in tree protection, and neighbours working on lake restoration in my home city of research.

Working on the urban commons with neighbours who became close friends, we argued and debated and discovered surprising similarities, and differences, in how we envisioned a good life, our desire for development and modernity vs culture and heritage, priorities for nature, biodiversity, and politics. Matches and mismatches in our norms and values, and in our politics—right and left, green and red—shaped how we fought, argued, and set aside deep divides to focus on common goals, in the process developing a deep respect for each other.

Norms, values, relational goals and ethics are at the core of how each one of us—academics, practitioners, policy makers and actors on the commons stage—engage with ideas of commons governance, and with commoning too. But rarely do we pay this important issue the attention it deserves.

The neglect of ethics is a reflection of a broader problem that underlies more than commons scholarship, indeed impacting the study of all aspects of sustainability—a western/global north bias that favors a narrative of 'value-neutral' research. Because of this western bias, much research on the commons (and other kinds of academic research) seeks to claim that the frameworks, theories, methods and even the cases we choose to select are agnostic to personal norms and ethics. Yet, as Stefan Partelow succinctly argues in the Introduction, our academic frameworks and ideas reflect our values, which in turn mirror our ethics—we would do well to acknowledge this openly, through honest conversations and debates that reflect the range of standpoints from which one can approach commons research ethically.

Through a series of chapters, this book discusses a range of different, yet interrelated aspects on ethics in commons research—focusing on justice, power, relational values, pluriversality and sacred relationships, in diverse contexts from a range of standpoints from the academic to the action researcher. By shining a light on many important aspects concerned with normative and ethical aspects of commons governance, these chapters provide a fresh, expansive and nuanced foundation for new research.

This critical book of commons scholarship aims to drive new conversations and open the door to fresh scholarship on ethics, and on how we view the commons and the role it plays in shaping a good life for all. It is a deeply important book for all those who study, work on, and value the commons.

Harini Nagendra
School of Climate Change
and Sustainability
Azim Premji University
Bangalore, India

Preface

In February 2023 I joined the Center for Life Ethics at the University of Bonn. Joining the CLE meant furthering my scholarly commitments to inter- and trans-disciplinary research even further, wading deeper into the depths away from being a specialist, and towards having a toolbox and bridging mindset. I have never considered myself as having a core disciplinary background, coming from environmental studies, sustainability science, and political science (due to the need to pick a discipline for doctoral studies on ocean and coastal commons). For the field of commons scholarship, interdisciplinarity and problem-focused research is already somewhat of the middle-ground. However, the link to ethics and ethical values has not always been clear or directly engaged with, even if implied within much of the field's classical and critical work. Although I am not a trained ethicist or philosopher, being at the Center for Life Ethics offers me an opportunity to bring the field of commons research more directly in contact with ethics and philosophy. The opportunity to begin building this bridge was the core inspiration for this book and the ongoing engagement with the claims and visions within it. Bridge building is both symbolic and practical, often connecting things that have not been connected before is now a foundational way in my research for generating new insights and approaches premised on inter- and trans-disciplinary ideals to tackle sustainability problems. I have experienced a lot of potential in collaborative research that can help us re-think our assumptions, sometimes allowing us to be reflective—both critically and constructively—for the first time or in new ways. The ability to see ourselves, our biases, and limitations—and our ethical values (and the value and influence of our ethical values in our research)—from a neutral or outside perspective. In my view, some of the strongest potential for scientific advancement and innovation resides in these boundary spaces. In many ways, because much of the knowledge we need is already available, it just needs to be considered together—in combination—and through dialogue. This, however, is not always easy. It requires willingness, patience, openness, and curiosity. In my view, it also requires the ability to validate and pursue new ideas—with both intrigue and heathy criticism—when they arise. All too often I get the impression that many of us have innovative thoughts and ideas, but lack the support, confidence, or spaces to discuss them openly—without the heavy weights attaching us to our identity.

Many ideas may not lead anywhere. We need not be attached to whether they lead somewhere or not, but rather to the process of engaging with them curiously and scientifically for the potential benefit of others. I believe we can fruitfully engage with important moral questions in this way. With these thoughts, the idea for this book emerged. I look forward to continued engagement and discussions about its content and chapters, and the inspiration that will come from building more bridges.

Bonn, Germany
April 2025

Stefan Partelow

Acknowledgements

This book would not have been possible without the support from the Center for Life Ethics and the University of Bonn. I consider it a privilege to work at both organizations, and particularly the Center for Life Ethics with its highly supportive and kind colleagues and leadership. Coming to the Center has deepened my knowledge and interested in ethics and its link to my research on the commons and environmental governance, which has inspired this book. I would like to thank the chapter contributors and peer-reviewers for the extensive work to develop the chapters, improve them, and engage in a collective process. Furthermore, numerous chapters draw on empirical research and direct experience within communities of practice who have given their time and insights to contribute to the knowledge within this book. Without these rich empirical links, the knowledge within this book would not be possible. Finally, I would like to acknowledge the many colleagues and collaborators that have provided inspiring content discussions and reflections for this book. Stefan Partelow is supported by the European Research Council (ERC) under the European Union's Horizon Europe research and innovation programme (Grant Agreement No. 101164978) for the ERC Starting Grant project SharedSeas.

Competing interests The editor and authors declare no conflicts of interest.

Ethics approval All necessary ethics procedures and approvals were obtained and followed. Specific procedures are documented in each respective chapter as needed.

Contents

Editor and Contributors

About the Editor

Dr. Stefan Partelow is a Professor of Environmental Governance at the University of Bonn in Bonn, Germany. He leads the Governance and Transformation group at the Center for Life Ethics, and is the Principal Investigator of the European Research Council (ERC) Starting Grant project called SharedSeas (2025-2029), which examines how to scale effective governance in our oceans. Dr. Partelow is a Lead Editor at the journal People and Nature, and the co-host and co-founder of the In Common Podcast, the official podcast of the International Association for the Study of the Commons (IASC). His research areas include ocean governance, small-scale fisheries and aquaculture, social-ecological systems analysis, commons, commoning, and institutional change. He completed his Doctoral and Postdoctoral research at the Leibniz Centre for Tropical Marine Research (ZMT) and Jacobs University (now Constructor University) in Bremen, Germany. Dr. Partelow further studied Environmental Studies and Sustainability Science at Lund University, Sweden, and the University of California, Santa Barbara.

Contributors

Liana O. Anderson Instituto Nacional de Pesquisas Espaciais, São José, Brazil

Evan J. Andrews Geography, Memorial University of Newfoundland, NL, Canada

Cécile Barnaud Dynafor, INRAE, Toulouse University, Toulouse, France

Maria B. Battaglia School of Environment, Resources and Sustainability, University of Waterloo, Waterloo, Canada

Guido Caniglia University of Helsinki, Helisinki, Finland

Mauricio Castrejón Universidad de las Américas-Programa Galápagos, Quito, Ecuador

Floriane Clement Dynafor, INRAE, Toulouse University, Toulouse, France

Erin Rizzato Devlin University of the Highlands and Islands, Inverness, Scotland, UK

Graham Epstein Blue Range Labs and School of Environment, Resources and Sustainability, University of Waterloo, Waterloo, Canada

Chris Giotitsas Ragnar Nurkse Department of Innovation and Governance, Tallinn University of Technology (TalTech), Tallinn, Estonia

Svein Jentoft Norwegian College of Fishery Science, UiT - The Arctic University of Norway, Tromsø, Norway

Isabel Jorgensen School of Environment, Resources and Sustainability, University of Waterloo, Waterloo, Canada

Ella Kari-Muhl School of Environment, Resources and Sustainability, University of Waterloo, Waterloo, Canada

Vasilis Kostakis Ragnar Nurkse Department of Innovation and Governance, Tallinn University of Technology (TalTech), Tallinn, Estonia;
Berkman Klein Center for Internet & Society, Harvard University, MA, USA

Asimina Kouvara Ragnar Nurkse Department of Innovation and Governance, Tallinn University of Technology, Tallinn, Estonia

Rafaella Lobo Nicholas School of the Environment, Duke University, North Carolina, USA

Christopher Luederitz Department of Management, University of Quebec at Trois-Rivières, Trois-Rivières, Canada

Guilherme Marques Institute of Hydraulic Research, Porto Alegre, Brazil

Stefan Partelow Center for Life Ethics, University of Bonn, Bonn, Germany;
Faculty of Agricultural, Nutritional and Engineering Sciences (AEI), University of Bonn, Bonn, Germany

Lily House Peters Geography, California State University, California, USA

Jeremy Pittman School of Planning, Faculty of Environment, University of Waterloo, Waterloo, Canada

Christina Priavolou Amsterdam Institute for Social Science Research, University of Amsterdam, Amsterdam, Netherlands

Pisith San Ragnar Nurkse Department of Innovation and Governance, Tallinn University of Technology, Tallinn, Estonia

Helena Shilomboleni School of Environment, Resources and Sustainability, University of Waterloo, Waterloo, Canada

Micaela Trimble Universidad de la República & South American Institute for Resilience and Sustainability Studies, Montevideo, Uruguay

Christiane Woopen Center for Life Ethics, University of Bonn, Bonn, Germany

Gabriela Alonso Yáñez Werklund School of Education, Specialization, Curriculum and Learning, University of Calgary, Calgary, Canada

List of Figures

Many Worlds, One Earth: Towards a Diverse Ethics of the Commons

Ethical Commoning: A Framework for Identifying and Comparing Values, Practices, and Capacities

A Look at the Commons Through the Lens of Buddhist Ethics

Knowledge Ethics for Marine Spatial Planning: Balancing Critical and Constructive Engagement with Small-Scale Fisheries

Trade-Offs Conducting Action Research on the Commons, Balancing the Ethics of Research and Practice

[illegible]

List of Tables

Ethics and the Commons: An Introduction to Why Building Ethical Awareness Matters

Justice and the Commons: A Review

Ethical Commoning: A Framework for Identifying and Comparing Values, Practices, and Capacities

Navigating Ethical Spaces in Common Property Research

Ethical Dimensions and Practical Guidance for Transdisciplinary Research on the Commons

Commons as Boundary Objects for Ethical Awareness: Key Questions and Directions Forward

Ethics and the Commons: An Introduction to Why Building Ethical Awareness Matters

Stefan Partelow

Abstract Ethical values are involved in every intentional human action—both in research, governance and civil society. The questions of this book are: what ethical values, challenges, and actions are associated with the commons and commoning? Do commons have a specific ethical and moral orientation? How can we engage with different and plural orientations? While the ethical values of commons may be implied from our research concepts, frameworks and methods, they are rarely made explicit or linked to ethical claims or theories in philosophy. This book aims to build awareness around the different ethical values, positions, and perspectives linked to commons, common property, and commoning research and practice. Building ethical awareness can help scholars strengthen their tools and skills for analyzing commons. Improving ethical awareness can give scholars a range of philosophical arguments to navigate and justify links between ethical values, actions, and outcomes. Current literature on the commons offers a diversity of concepts and arguments related to the benefits and challenges of commons, common property and commoning for the governance and sustainability of social-ecological systems. One challenge is that commons are not one thing. The term 'commons' is broadly used to represent different objects of shared interest, types of governance, social actions, and ideological positions. Given the diversity of commons and their tendency to have strong place-based dynamics, a plurality of ethical orientations and claims are observed and perhaps expected in the literature. Acknowledging and working with multiple ethical orientations, theories, and perspectives can enable richer analytical and reflective perspectives that can improve our engagement with the ethical complexities of researching and governing commons. However, differentiating current ethical orientations in the literature can be challenging. This introductory chapter explores different arguments and perspectives for interpreting the commons, common property and commoning literature. To conclude, the organization of the book is briefly introduced, including the final chapter which provides key questions for the field

S. Partelow (✉)
Center for Life Ethics, University of Bonn, Bonn, Germany
e-mail: stefan.partelow@uni-bonn.de

Faculty of Agricultural, Nutritional and Engineering Sciences (AEI), University of Bonn, Bonn, Germany

© The Author(s) 2026

S. Partelow (ed.), *Ethics and the Commons*, Schriften zu Leben und Gesellschaft - Series on Life and Society 3, https://doi.org/10.1007/978-3-032-24246-4_1

going forward and argues that commons can act as boundary objects for building ethical awareness.

1 Introduction

I would like to argue that the terms commons, common property, and commoning are unavoidably attached to sets of ethical values. This may seem obvious, but the claim is discussed surprisingly little in the field's literature. Maybe this is because of how science—in its contemporary Western-origin form—largely ignores its link to values and normative positioning, or claims itself to be objective in its approach to understanding reality, and therefore value-neutral. The evolution of the field of commons research over time helps us understand this. Despite the field's current interdisciplinary nature, its history is firmly rooted in public administration and institutional economics (Ostrom, 1998; Partelow et al., 2020). Interdisciplinarity on the commons still lacks intellectual exchange with fields such as philosophy, arts, and the humanities (van Laerhoven et al., 2020), which engage with topics of value claims and their representation more frequently. Engagement with these dimensions is important because all human actions, such as the creation of concepts and narratives, are linked to ethical values. Concepts and narratives reflect our values and are powerfully influential in science. For example, the way we conceptualize an object or institution creates an imagined reality about them (Jentoft, 2023), fulfilling our human need to clarify the often-intangible uncertainty and complexity of reality, but also the scientific need for orientation in our proposals, faculties, and communities. Our academic concepts (i.e., terms, frameworks, theories) reflect our values, and our values mirror our ethics. They are like bridges between what is real and the actions we take in relation to them. They are our filters for making sense of complexity and decision-making amidst what Ostrom may have called our bounded rationality. They are ethical—and because they are ethical—they matter for how we act in relation to them and because of them.

For the field of commons research, the terminology of commons, common property, and commoning are our central unifying concepts. They help us orient to each other and guide our actions. When I see a paper using these terms, I somewhat unconsciously assume there are unifying features or similarities in between myself and the authors, such as the questions we might ask (i.e., epistemic positions). These terminologies are our commons—our orienting lighthouses on the academic seascape—they are what we share as members of a research community. I would argue that our terminologies, frameworks, and concepts are even boundary objects. Boundary objects are interpretatively flexible, have emergent structures, and have multi-purpose adaptable functions (Leigh Star, 2010). Most importantly, boundary objects sit between things, and because they sit in the middle, they act as important bridges and translocators of ideas, values, and ethics.

Yet, we struggle to define commons specifically—amidst their plurality—particularly their ethics (Mandalaki & Fotaki, 2020). It is hard to grasp and clearly define malleable objects—like commons—because they shape-shift to fit where they are. They also shape-shift to fit what we want to believe about them. Due to this, one might get the impression that specifying the ethics of the commons is an actively avoided scientific endeavor—a perceived relative pursuit of moral indifference (e.g., place-specific contexts matter to the point that comparison is not important). Yet, our use of concepts implicitly related to ethical claims seems to be increasing, such as with justice, equity, gender, conservation, or sustainability (van Laerhoven et al., 2020). It is becoming more and more difficult to be certain about whether our assumptions and interpretations about commons are descriptive or normative—inductively examined yet predetermined in how we will see them. Am I neutrally observing ethical actions? Or am I wearing thick glasses that filter what I want to see? We are good at seeing what we want (but often not aware of it), and if we look for something, we will likely find it somewhere. In prior research reviewing different environmental governance theories (Partelow et al., 2020), we argue for a toolbox approach, because of this hammer-nail problem.[1] Any specific value or governance approach may be partly identifiable in most systems, but is this what we want to be there? A toolbox of concepts and approaches can better help us analyze, compare and contrast what we observe in our science to better unpack value questions about what is viable, desirable, and feasible. We could likely find evidence that partially or even fully validate many theories or concepts. This tells us a few useful things. Either our theories are incomplete representations of a more complex multi-sided or plural reality, and to understand that reality we need to improve our theories or combine them. Or, multiple realities are possible, and we have a choice about what we want to know and show, and how to get there. Here, our ethics matter.

To be clear, the chapters in this book do not accumulate into to a single definitive conclusion about what the ethics of the commons are. It is not clear that any singular ethical position could be identified among the plural literature, nor would it (likely) be desirable to have one. Instead, what the field needs, and what we argue as a collective author group in the final chapter of this book, is to build more ethical awareness around the commons (and other fields of science and sustainability)—to expand our analytical and reflective set of tools and skills—to improve our analytical capacities about commons, and the actions we take as a result.

[1] When you have a hammer, everything looks like a nail. The so-called Law of the Instrument refers to using the tool you have for all purposes, even if it is not fit for use. In science, when you have a theory you like or want to test, you will look for observations in support or against it, and likely find evidence in favor while ignoring other tools to examine those observations as a form of confirmation bias. There may be evidence supporting many theories. The question is, what is a useful means, and towards what end?

2 Defining the Commons

Here is a basic but tricky question: what are the commons? Mining the literature and online sources quickly reveals a wide array of interpretations. I believe the field would benefit from developing clearer conceptual frameworks for parsing out more detailed axes of differentiation between resources, rights regimes, institutional bundles, problem frames, and ideologies that broadly encompass a wider variety of different institutional and human-nature configurations. The current major axes are either commons as resources with inherent properties that position them as either common-pool resources or public goods (i.e., based on degree of rivalry and excludability). Linked to this, is the notion that commons are open to all, and are the resource we must share. This framing contrasts with the next framing of commons as a rights regime, which are not open to all, but require clear social and ecological boundaries to govern. This type of common property requires governance regimes generally based on the need for collective action, cooperation, and coordination, whether self-organized or externally imposed. The main confusion here is between purposeful commons property regimes versus open access regimes (also often referred to as commons) with no clear rights, institutional configurations, or group responsibilities. A key difference being, that when clearly defined groups have common property rights to place-based commons over time, deep human-nature relations often form in tangent with local knowledge and experience-based rules and norms for collective use.

Another narrative/conceptual tension is that commons are situated between a problem frame and a solution option. For example, common pool resources and public goods are both shared resources with inherent problem properties (e.g., subtractability, excludability), but commons as a rights regime are a solution, and sometimes these rights regimes are attached to normative ideologies (e.g., community justice, degrowth, collectivism, livelihoods)—but sometimes—also attached to descriptive observations of functional institutions for collective action in natural resource governance. This plurality can be confusing. Not necessarily problematic when transparently clarified (although few papers do this), but confusing because a handful of similar but different narratives are woven together into the illusion of a singular object 'the commons', which in my view positively embodies plural forms, institutions, ethics, and even plural ontologies.

Furthermore, there is the verb commoning, which has emerged to examine sets of processes and actions among groups sharing resources or institutions to solve problems together as communities and to develop governance institutions collectively. Commoning seems to represent a shift in what the intended goal of effective commons governance is, at least among scholars who observe this in practice communities. Classically, this was to sustain the use of the biophysical commons (i.e., fish, trees, water), which is where the commons literature has originally intersected with the sustainability and social-ecological systems literature. Commoning represents a narrative shift towards valuing the social community and community values of collective action itself, as the resource we want to be sustained. Here the claims tend to revolve around being or becoming commoners through commoning practices, which exist on the spectrum from building social movements of resistance as the primary orientation (e.g., anti-colonial, anti-capitalist, degrowth) to positive visioning of future societies (e.g., sense of togetherness, care, solidarity, more-than-human, urban collectivism). Highlighting these observed differences helps illustrate the plurality of ethical values that underpin different dimensions of what we know to be the commons, the researchers who study them, and the practices/actions embodied in them.

An additional layer worth mentioning, is where our field's meta-narratives loom, which are additionally reflective of ethical values. The 'tragedy' narrative makes assumptions about human actions (or lack of actions). These include that people sharing resources would not talk, that they would view resource commons instrumentally (an ethical assumption), that they would have similar intentions, values, or risks, and that they would pursue non-adjustable self-interested behaviors isolated from their social-ecological contexts and relations. However, a single narrative cannot capture what we now know is a pluriverse of realities that range from verifying elements of the tragedy claim to completely dismissing it. The Ostroms and colleagues later course corrected some of the claims of the tragedy narrative with what could be interpreted as a technical management of complexity approach through institutional development and change (a critique of this is explored further below), and a lot of progress was made. Large gaps in scientific and local/traditional knowledge synthesis were filled in understanding the viability and feasibility of sustainable community-based natural resource management and common property regimes, validating countless indigenous and local community-based practices as more effective than (the often) coercive state (and often colonial) oversight that lacked contextual nuance and care for local culture, context, and practices. Notable books like Seeing Like a State (Scott, 1998), Environmentality (Agrawal, 2005), and Common Boundaries (Cox, 2024) have illustrated these issues in detail. State oversight is not something that can be removed in most present-day contexts, but the relationship with local natural resource user communities can be governed in many different ways. Shifts towards co-management or collaborative governance have emerged in many countries to govern natural resources with local users as a result. This trend has been championed as necessary progress for two reasons that are a mix of objective observations of commons sustainability outcomes and normative in that collaborative governance has more opportunity for inclusion and participation for local communities and stakeholders, although the reality of co-management in practice is often associated with struggles over long time periods for power, knowledge validation, and capacity investments into and from local communities facing risks (Murunga et al., 2021; Partelow et al., 2018; Senghor et al., 2023). Contextual nuances, complex system understandings, and local goals often now take center stage in framing the need for place-based management and governance, but they have arguably been dominated by social-technical approaches and Western-based administrative thinking and science which often overlooks aspects such as systemic injustices, different worldviews, power imbalances, and alternative ways of knowing and being.

At the core of commons research—beyond framings and narratives—are the gritty details of what our research looks like on a day-to-day basis. We spend a lot of time discussing frameworks. How to interpret them, define, explain them to students, all the while justifying their existence while being consistently puzzled about how to do empirical research with (operationalize) them with data collection, analysis, and comparative methods. We need orientation in our academic communities, and frameworks are the centers of gravity (at least in the environmental social sciences) that provide that orientation (Cox et al., 2021; Partelow, 2023). In our papers we use frameworks to identify the social, institutional, and biophysical features of place-based commons, and how those features relate to processes and outcomes for both people and nature. The field's numerous frameworks are constructed of concepts used to guide the questions we ask, data collection, which is used to help us analyze and build theory, so the story goes. In practice, it is messier. We ask certain questions based on personal curiosity and what we think might be novel or publishable or fundable, we have career ambitions, we want to work with specific communities or colleagues and align the approaches we take in science with our lives inside and outside of professional work. We are path dependent in our thinking and practices, and go back to work with the same communities and the same colleagues in the languages that are familiar. We often want to practically help those places alongside the research running efficiently, while much of the empirical work is done by first-time early career individuals given time constraints, teaching obligations, and educational/supervision needs. All these things are completely reasonable things to do, but they are the reality for most professionals in contemporary science systems. But, nonetheless these are things that are guided by ethical values that co-mingle, intersect, and perhaps conflict. This is not a negative portrayal or criticism of science systems, nor is it unique to the field of commons research. Science—as a system of plural knowledge production institutions and values—needs to be uniformly and regularly defended as one of civilizations most valuable spaces for independent thinking and knowledge creation—with near unimaginable contributions supporting human well-being and flourishing. Recognition of the role of our ethical values in science is rather a call to be cognizant and transparent that ethics run-through our actions, both in how we do science and the outcomes of knowledge products that emerge.

In doing so, the commons field has identified many features that recur across cases, and which have become central tenets of what we understand and believe about commons (McGinnis, 2011; Ostrom, 1990). The ethical claims of these features are rarely discussed and often not clear, neither are the ethics of balancing what is place-based versus broadly applicable or transferrable, or desirable; and from whose perspective? A final and perhaps more surface level observation that reflects more rooted epistemic intentions is warranted. The content of the frameworks themselves—e.g., bundles of rights, Design Principles, SESF,[2] IADF[3]—don't include explicit ethical values or even mention ethically related features such as virtues, maximizing, sufficiency, duties, or even concepts like power, agreement, or justice

[2] Social-Ecological Systems Framework (McGinnis & Ostrom, 2014; Partelow, 2018).

[3] Institutional Analysis and Development Framework (McGinnis, 2011).

(Cleaver & De Koning, 2015). Yet, many are implicitly embedded. I must admit, when commons scholars are asked to imagine something, a natural skepticism arises. But if we imagine the iceberg model of the commons field—where our concepts, terms and frameworks are sticking out of the water—what we don't see are the intentions and ethical values below as easily.

3 Ethics and Values in Current Commons Scholarship

Although rather minimal, there has certainly been debate and publications on ethical issues in the literature. This has largely emerged from what is referred to as critical commons scholarship because it challenges the assumptions and outcomes of the new institutionalist approaches that popularized the field (Fig. 1). The term critical here means more investigative and reflective on the potential deeper assumptions, meanings, oversights, imbalances, or biases of the scholarship on the commons that emerged from public administration and institutional economics. Quintana and Campbell (2019) reviewed the critical commons literature and argue there are roughly four types: Type 1 is the functionalist critique which argues that focusing narrowly on institutions neglects history and context and therefore misrepresents causes of sustainability outcomes. These arguments have largely been made by critical institutionalists, that there is more to commons then their institutional configurations (Campbell et al., 2001; Cleaver, 2000; Cleaver & De Koning, 2015; Saunders, 2014). Type 2 is the apolitical management critique which argues that a focus on resource sustainability causes commons scholars to ignore how social issues of power create and maintain inequalities in access (Agrawal, 2003; Mosse, 1997; Ribot & Peluso, 2003). Type 3 is the methodological critique which argues that methodological incompatibilities don't allow the field to move beyond general models and theory (Johnson, 2004), in a way that sufficiently allows them to address the Type 1 and Type 2 issues. Type 4 argues that current scholarship is used to support neoliberal practices and power imbalances (Agrawal, 2005; Mansfield, 2004). It is clear the ethical values are a part of each type, and although not explicitly discussed as ethics, they initiate discussions towards building ethical awareness around intentions and actions.

Most of the research that connects commons with philosophy, political values, or ethics comes from activists, writers and scholars outside of academic—sometimes with academic partners—such as the work of David Bollier and Silke Hilfrich (Bollier, 2020; Bollier & Hilfrich, 2012, 2015), or Jay Walljasper (Walljasper, 2010), and of course many others including countless non-Western advocates and communities. These works remind us that the commons are linked to real and diverse social movements and practice communities enduring political and cultural struggle. As noted by numerous practitioners, the commons represent a network of movements across societies worldwide to take actions toward alternative futures of social organization and values, within what Bollier (2024) calls the Commonsverse.

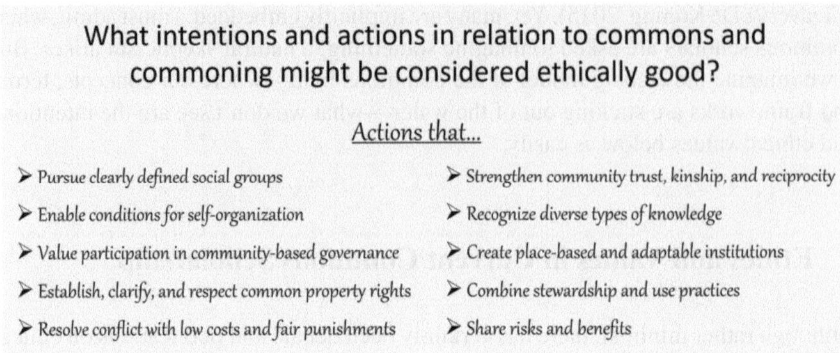

Fig. 1 Do the commons and commoning have explicit ethical values and value practices? Here are some of the well-accepted features and conditions of commons that are successfully governed. Are they a basis for an ethics of the commons? An important connection for building ethical awareness is to assess these claims alongside the ethical theories and practices listed in Table 2. The list in this figure is not comprehensive or based on analysis, but aims to inspire the question and ideas being asked. The figure only depicts text as a list of intentions and actions relating to the commons that might be considered ethically good

From academia, a recent article by Perperidis (2024) argues that a good portion of the ethical values of the commons are rooted in political dimensions opposing neoliberal governance practices such as privatization and individualism, and commons are therefore linked to acts of political and economic organization strategies, often via self-organized social movements. These arguments are reflected in examples of movements such as rights-to-the-city, peer-to-peer (P2P) production, or economic degrowth narratives. A counter argument to this would be that this is an interpretation of the commons that has come out of Western or European thinking, and does not account for the plurality of commons ethics from indigenous or rural non-Western communities (see the ethical cultural practices in Table 2). Economic arguments also overlook arguments that the commons are fundamentally a question of ethics, not economics (Boda, 2003). Perperidis (2024) further highlights that the commons are linked directly to environmental ethics. However, in Western science and public discourse, particularly conservation science, our narrative framing of environmental ethics has changed over time and do not really have a central unifying claim. Mace (2014) argues that Western narratives of people and nature relationships—underpinned and reflected in conservation science—have had four phases: nature for itself (1960–1970s), nature despite people (1980–1990s), nature for people (1990–2000s) and people and nature (2010–2020s). There is one clear divide between mainstream nature conservation approaches and commons-based approaches that can be highlighted. Commons are almost always an integrated people and nature model, where ethical values and actions for the stewardship and use of resources emerge from/drive interwoven actions, where people can't be separated from nature. Although changing in recent decades, Western conservation often excludes people, and severs this linkage (Brockington, 2002; West et al., 2006), and in doing so also cuts the local values and cultures people have to land, sea, and resources, mandating that people either move or source their resources from places where they have no connection or option to build direct stewardship values. As such, environmental ethics from Western conservation is ontologically distinct in that people have ethics towards nature, rather than people being intricately inseparable from nature ontologically.

Further commons literature has examined more specific topics or concepts related to values and ethics (Fig. 1). Murillo et al. (2024) examine the ethics of commons organizing and put forward a handful of key features of ethical relevance when doing so, including: excluding, subtracting, type of usage, polycentricity and subsidiarity, individual engagement, community, human character, and environmental stewardship. Commons organizing ethics are further explored by Albareda and Sison (2020). These dimensions are individually unpacked and put together in a framework that also examines the modern challenges of what the authors call normative commons organizing, including new forms of alienation, commodification and the lack of accountability. Furthermore, Périvier (2024) reviews literature from gender studies and the commons. The review explores the links of gender studies to feminist movements and related political agendas. Overall, the study notes the importance of cross-sectional approaches to examine the intersection of gender and the commons in the context of three societal transitions—environmental, demographic, and digital. Fuchs (2020) examines the ethics of digital commons, for example, from a virtue ethics perspective. Partelow and Manlosa (2023) review the commoning literature and how it links to power dynamics through an analytical framework linked to coercive power over, collaborative power with, agentic power to, psychic power within (Chambers, 2006; VeneKlasen & Miller, 2007). Furthermore, Mandalaki and Fotaki (2020) draw on feminist ideas to explore a relational embodied ethics of the commons linked to commoning and social organization processes, exploring the details of vulnerability, reciprocity, and embodied relationality at the linked individual and communal levels.

4 Building Ethical Awareness: What Tools and Skills Do We Need?

Ethics refers to the orienting moral principles, attitudes, or goals that guide actions. These are criteria with which people, an action, an institution, or a system of rules or norms are ethically examined or evaluated. As such, ethical values are deeply integrated with the processes and goals of commons, commoning, common property governance, and institutional development and change (i.e., rules, norms, and social organization and decision-making structures). These definitions fit both the evaluation of our research ethics, and the ethics of commons as objects in society for evaluation through research and practice. There is a need to build ethical awareness of different ethical values and how to evaluate them, and to do so in specific research areas for appropriate fit such as within knowledge co-production approaches (Partelow et al., 2024). Ethical awareness can strengthen how and why we justify and argue the potential benefits and challenges of the commons by providing a toolbox of theories, concepts, and methodologies for reference and use. A toolbox is defined as a collection of practical resources and skills including key concepts, theories, principles, and methods designed to assist researchers in navigating, in this case, ethical values and moral dilemmas (Baggini & Fosl, 2014). In Table 1, I attempt to categorize and specify the types of tools and skills that—as individuals, working groups, and scientific institutions—might improve our ethical awareness.

Knowledge of different ethical theories is an important knowledge tool that provides insights into the century—if not millennia—long philosophical writings and debates across history, which remain cornerstones of scholarly research on ethics today. There are many ethical theories that have emerged from different cultural practices, scientific fields and philosophical debates (Table 2). In this chapter, I highlight a limited number of theories to exemplify the diversity of ethical claims that exist and are influential in the current literature and discourse. I recognize the potential

Table 1 Tools, skills, and investments helpful for building ethical awareness. There are three broad categories, within which there are specific tools, skills, and investments. This is preliminary list, and is not comprehensive or derived from a systematic assessment

Categories	Specific tools and skills
Knowledge content	• Awareness of different ethical theories and their claims (see Table 2) • The ability to engage with and differentiate descriptive and normative ethics • The ability to provide clear definitions of concepts and terminology • The ability to integrate theory and empirical research • The ability to be critically reflective of underlying claims and assumptions • The ability to orient and position your research and the research of others
Knowledge governance	• Investments into funding and support for diverse collaborative research and dialogue • Investments into funding and support for place-based ethics and community engagement • Access to and engagement with ethical review boards and committees • Investments into processes and safeguards for data storage, data anonymization, and re-use • Investments into processes for institutional reflection, learning, and change
Capacity-building	• Investments into education and training programs that support engagement with ethics • The ability to listen, be present, critically reflect and constructively engage with others • The ability to participate in and/or lead collaborate research • Investments into self-care, mentorship, and coaching within teams and institutes • The ability to engage with and effectively communicate with non-academic communities

Table 2 Here is a list of simplified and interpreted claims of different ethical theories and broader cultural approaches to ethics originating from different time periods and world regions. When thinking about the commons in the context of each theory or approach, recognizable features and terminology can be observed as being associated with different aspects of the commons and commoning literature. The claims are a brief interpretation of each approach, and do not intend to represent the complexity and depth that each approach embodies. It is not our intention to misrepresent or misuse the ethical theories or cultural approaches in any way. In the spirit of building ethical awareness, we believe it is important to acknowledge the many different cultural practices and their contributions to ethical thinking beyond Western/Northern academic theorizing. This table was derived for the final chapter of this book, and subsequently added to this introduction for reference, recognizing the contributions of all authors contributing to the final chapter

Ethics approach	Theory/ Practice	Origin/Notable figures	Simplified ethical claims
Discourse ethics	Specific theory	Habermas (Germany) Apel (Germany)	Actions are ethically good when … they result from a discourse following defined rules of equality, respect and fair participation of all persons concerned
Virtue ethics	Specific theory	Confucius (~500 BE, China) Plato (ancient Greece) Aristotle (ancient Greece)	Actions are ethically good when… they arise from virtues such as generosity, justice or compassion. Ethically good actions are those that a virtuous person does
Consequentialism	Specific theory	Bentham (UK) Mill (UK)	Actions are ethically good when… their consequences maximize well-being and minimize harm for the greatest number
Feminist ethics	Specific theory	1970s (USA/Europe)	Actions are ethically good when… they address power imbalances, support relational care, reorganize reproductive labor, and promote inclusivity and equity
Contractualism	Specific theory	Hobbes (UK) Rawls (USA) Scanlon (USA)	Actions are ethically good when… they emerge from agreements that are justifiable to all affected parties, ensuring fairness, respect agency, and mutual benefit
Deontology	Specific theory	Kant (Germany)	Actions are ethically good when… they are guided by maxims which can be universalized according to the categorical imperative. Moral duties apply to everyone such as respecting human dignity and autonomy

(continued)

Table 2 (continued)

Ethics approach	Theory/ Practice	Origin/Notable figures	Simplified ethical claims
Life ethics	Specific theory	Center for Life Ethics (Germany)	Actions are ethically good when… respecting or fostering the unfolding, plenitude, beauty, and embeddedness of all life.
Environmental ethics	Specific theory	Carson (USA) Leopold (USA) Naess (Norway)	Actions are ethically good when…moral consideration extends beyond humans to nature and more-than-human entities to recognize the interdependencies and relational connections between the biological community
Ubuntu ethics	Broad cultural practice	Indigenous Peoples of Southern Africa	Actions are ethically good when… they promote interconnectedness, mutual care, and the well-being of all in the community, reflecting "I am because we are"
African Palaver ethics	Broad cultural practice	Indigenous Peoples of Sub-Saharan Africa	Actions are ethically good when… they originate from a consensus that is reached by dialogue in spaces for open communication for all persons of the community
Buddhist ethics	Broad cultural practice	~500 BCE (India)	Actions are ethically good when… they originate from infinite compassion that is free of desire and are aligned to purity and perfection of mind. This entails awareness of interdependence with all life
Buen Vivir	Broad cultural practice	Andean Indigenous Peoples (South America)	Actions are ethically good when… actions pursue reciprocity and care, harmony with nature, collective wellbeing, sufficiency, plurality, and relationality
Seven Grandfather Teachings	Broad cultural practice	Anishinaabe Indigenous Peoples (North America)	Actions are ethically good when… they are guided by the principles of love, respect, bravery, truth, honesty, humility, and wisdom

biases in choosing certain ethical theories over others, and the selection of the theories in Table 2 is by no means comprehensive, representative, or based on a detailed analysis of each theory or practice. Nonetheless, they provide useful contrasts to each other, and represent approaches that are frequently engaged with from different world regions, time periods, and epistemic orientations.

Another important skill to strive for, is the ability to differentiate and position the research we do (including contexts examined, terms, and concepts used) in the field and in relation to others. Why is this important? Positioning and orientation are, in part, the ability to understand the intentions of the research, which reflect the ethical values behind it, or the ethical issues that may (or might have) arose during the research itself. I find it useful to use multiple axes of differentiation as orienting tools. Axes of differentiation refers to thinking about the spectrum between two concepts that usually have opposite or oppositional components to their meanings. Useful axes include: descriptive versus normative, process versus outcome, place-based versus generalizable, or positivist versus constructivist ontology. There are of course other useful orientations such as disciplinary, methodological, or cultural (both academic and empirical). Engaging in exercises to position our research is a useful step forward towards realizing the plurality of positions and ethical values driving research beyond our own, and to help influence our curiosity about other approaches, and in turn, force ourselves to reflect on how we justify our approaches in relation to others.

5 How to Engage with This Book

The chapters in this book offer a range of arguments and contexts for building ethical awareness surrounding commons, common property and commoning topics. In my view, a useful way to approach the individual chapters is to orient your own research and perspectives to them, perhaps reflecting on the tools introduced above—or your own tools—to discuss them with colleagues. The chapters offer a range of detailed accounts on specific topics, but also clear takeaways that have linkages beyond their contexts. More specifically, the first three chapter offer reviews and perspectives on specific topics that include justice scholarship on the commons (chapter "Justice and the Commons: A Review"), pluriversality or many worlds approaches (chapter "Many Worlds, One Earth: Towards a Diverse Ethics of the Commons"), and the development of an ethical commoning framework (chapter "Ethical Commoning: A Framework for Identifying and Comparing Values, Practices, and Capacities"). After this, specific case study approaches are detailed which includes the analysis of Buddhist commons (chapter "A Look at the Commons Through the Lens of Buddhist Ethics"), marine spatial planning and transdisciplinary small-scale fisheries research (chapter "Knowledge Ethics for Marine Spatial Planning: Balancing Critical and Constructive Engagement with Small-Scale Fisheries"), and action research on the commons in the context of peer-to-peer production (chapter "Trade-Offs Conducting Action Research on the Commons: Balancing the Ethics of Research and Practice").

Each of these context specific case studies offer diverse insights into the wide range of ethical values and practices that emerge in both research and practice. Theses reflections are followed by two chapters that examine research ethics or research integrity issues in commons and commoning research, including examining the diverse ethical perspectives on how we do commons research (chapter "Navigating Ethical Spaces in Common Property Research"), and practices lessons from a range of transdisciplinary projects on the commons (chapter "Ethical Dimensions and Practical Guidance for Transdisciplinary Research on the Commons"). In combination, the chapters offer a mix of review and reflective insights combined with unique case examples and guidance for building awareness on research integrity on the commons. The final chapter (chapter "Commons as Boundary Objects for Ethical Awareness: Key Questions and Directions Forward") aims to provide conceptual and theoretical orientation (building on this introduction) to orient our researchers and practitioners in the field with specific ethical theories and ethically oriented cultural practices. The final chapter provide a range of specific tools and take-home guiding insights for building ethical awareness. The overall purpose of this book is to begin deeper and more comprehensive dialogue in our commons and commoning communities around ethics. This awareness building would include, in my view, improving how and why we use normative concepts in our research, our knowledge of ethical theories and claims in relation to our concepts and research questions, as well as the ability to critically but constructively engage with our own research assumptions, limitations, and actions within and between our academic and practice communities.

References

Agrawal, A. (2003). Sustainable governance of common-pool resources: Context, methods, and politics. *Annual Review of Anthropology, 32*(1), 243–262. https://doi.org/10.1146/annurev.anthro.32.061002.093112

Agrawal, A. (2005). *Environmentality: Technologies of government and the making of subjects.* Duke University Press.

Albareda, L., & Sison, A. J. G. (2020). Commons organizing: Embedding common good and institutions for collective action. Insights from Ethics and Economics. *Journal of Business Ethics, 166*(4), 727–743. https://doi.org/10.1007/s10551-020-04580-8

Baggini, J., & Fosl, P. S. (2014). *The ethics toolkit: A compendium of ethical concepts and methods.* Wiley-Blackwell. https://www.wiley.com/en-us/The+Ethics+Toolkit%3A+A+Compen...

Boda, Z. (2003). Global environmental commons and the need for ethics. *Society and Economy, 25*(2), 213–224.

Bollier, D. (2020). Commoning as a transformative social paradigm. *The New Systems Reader*, 348–361. https://doi.org/10.4324/9780367313401-28

Bollier, D. (2024). Challenges in expanding the commonsverse. *International Journal of the Commons, 18*(1), 288–301. https://doi.org/10.5334/ijc.1389

Bollier, D., & Helfrich, S. (2012). *The wealth of the commons: A world beyond market and state.* The Commons Strategy Group. https://wealthofthecommons.org/

Bollier, D., & Helfrich, S. (Eds.). (2015). *Patterns of commoning* (D. Bollier & S. Helfrich, Eds.). Commons Strategy Group and Off the Common Press. https://patternsofcommoning.org/

Brockington, D. (2002). *Fortress conservation: The preservation of the Mkomazi Game Reserve.* Indiana University Press.

Campbell, B., Mandondo, A., Nemarundwe, N., Sithole, B., & Matose, F. (2001). *Challenges to proponents of common property recource systems: Despairing Voices from the Social Forests of Zimbabwe* (Vol. 29, Issue 4). www.elsevier.com/locate/worlddev

Chambers, R. (2006). Transforming power: From zero-sum to win-win? *IDS Bulletin, 37*(6), 99–110.

Cleaver, F. (2000). Moral ecological rationality, institutions and the management of common property resources. *Development and Change, 31*(2), 361–383. https://doi.org/10.1111/1467-7660.00158

Cleaver, F., & De Koning, J. (2015). Furthering critical institutionalism. *International Journal of the Commons, 9*(1), 1–18. https://doi.org/10.18352/ijc.605

Cox, M. (2024). *Common boundaries: The theory and practice of environmental property.* Columbia University Press.

Cox, M., Gurney, G. G., Anderies, J. M., Coleman, E., Darling, E., Epstein, G., & Frey, U. J. (2021). Lessons learned from synthetic research projects based on the Ostrom Workshop frameworks. *Ecology & Society, 26*(1).

Fuchs, C. (2020). The ethics of the digital commons. *Journal of Media Ethics, 35*(2), 112–126. https://doi.org/10.1080/23736992.2020.1736077

Jentoft, S. (2023). The image fallacy: Rethinking the tragedy of the commons. *International Journal of the Commons, 17*(1), 174–183. https://doi.org/10.5334/ijc.1248

Johnson, C. (2004). Uncommon ground: The "Poverty of History" in common property discourse. *Development and Change, 35*(3), 407–434. https://doi.org/10.1111/j.1467-7660.2004.00359.x

Leigh Star, S. (2010). This is not a boundary object: Reflections on the origin of a concept. *Science, Technology & Human Values, 35*, 601–617. https://doi.org/10.1177/0162243910377624

Mace, G. M. (2014). Whose conservation? *Science, 345*(6204), 1558–1560. http://www.sciencemag.org/cgi/doi/10.1126/science.1254704

Mandalaki, E., & Fotaki, M. (2020). The bodies of the commons: Towards a relational embodied ethics of the commons. *Journal of Business Ethics, 166*(4), 745–760. https://doi.org/10.1007/s10551-020-04581-7

Mansfield, B. (2004). Neoliberalism in the oceans: "Rationalization", property rights, and the commons question. *Geoforum, 35*(3), 313–326. https://doi.org/10.1016/j.geoforum.2003.05.002

McGinnis, M. D. (2011). An introduction to IAD and the language of the Ostrom workshop: A simple guide to a complex framework. *Policy Studies Journal, 39*(1), 169–183. https://doi.org/10.1111/j.1541-0072.2010.00401.x

McGinnis, M. D., & Ostrom, E. (2014). Social-ecological system framework: Initial changes and continuing challenges. *Ecology and Society, 19*(2). https://doi.org/10.5751/ES-06387-190230

Mosse, D. (1997). The symbolic making of a common property resource: History, ecology and locality in a tank-irrigated landscape in South India. *Development and Change, 28*(3), 467–504. https://doi.org/10.1111/1467-7660.00051

Murillo, D., Guinart, P., & Arenas, D. (2024). The ethics of commons organizing: A critical reading. *Journal of Business Ethics.* https://doi.org/10.1007/s10551-024-05706-y

Murunga, M., Partelow, S., & Breckwoldt, A. (2021). Drivers of collective action and role of conflict in Kenyan fisheries. *World Development, 141.* https://doi.org/10.1016/j.worlddev.2021.105413

Ostrom, E. (1990). Governing the commons: The evolution of institutions for collective action. *Cambridge University Press.* https://doi.org/10.1017/CBO9780511807763

Ostrom, E. (1998). A behavioral approach to the rational choice theory of collective action. *American Political Science Review, 92*(1), 1–22. https://doi.org/10.2307/2585925

Partelow, S. (2018). A review of the social-ecological systems framework: Applications, methods, modifications, and challenges. *Ecology and Society, 23*(4), 36. https://doi.org/10.5751/ES-10594-230436

Partelow, S. (2023). What is a framework? Understanding their purpose, value, development and use. *Journal of Environmental Studies and Sciences*. https://doi.org/10.1007/s13412-023-00833-w

Partelow, S., Glaser, M., Solano Arce, S., Barboza, R. S. L., & Schlüter, A. (2018). Mangroves, fishers, and the struggle for adaptive comanagement: Applying the social-ecological systems framework to a marine extractive reserve (RESEX) in Brazil. *Ecology and Society*, *23*(3), art19. https://doi.org/10.5751/ES-10269-230319

Partelow, S., Luederitz, C., Huang, Y.-S., von Wehrden, H., & Woopen, C. (2024). Building ethical awareness to strengthen co-production for transformation. *Sustainability Science*. https://doi.org/10.1007/s11625-024-01582-7

Partelow, S., & Manlosa, A. O. (2023). Commoning the governance: A review of literature and the integration of power. *Sustainability Science*, 0123456789. https://doi.org/10.1007/s11625-022-01191-2

Partelow, S., Schlüter, A., Armitage, D., Bavinck, M., Carlisle, K., Gruby, R. L., Hornidge, A.-K., Le Tissier, M., Pittman, J. B., Song, A. M., Sousa, L. P., Văidianu, N., & Assche, K. Van. (2020). Environmental governance theories: A review and application to coastal systems. *Ecology & Society*, *25*(4). https://doi.org/10.5751/ES-12067-250419

Périvier, H. (2024). Crossing gender studies with the commons: A bibliometric analysis. *International Journal of the Commons*, *18*(1). https://doi.org/10.5334/ijc.1341

Perperidis, G. (2024). A philosophical examination of the ethical foundations of the commons. *International Journal of the Commons*, *18*(1), 276–287. https://doi.org/10.5334/ijc.1379

Quintana, A., & Campbell, L. M. (2019). Critical commons scholarship: A typology. *International Journal of the Commons*, *13*(2), 1112–1127. https://doi.org/10.5334/ijc.925

Ribot, J. C., & Peluso, N. L. (2003). A theory of access. *Rural Sociology*, *68*(2), 153–181. https://doi.org/10.1111/j.1549-0831.2003.tb00133.x

Saunders, F. P. (2014). The promise of common pool resource theory and the reality of commons projects. *Source: International Journal of the Commons*, *8*(2), 636–656. https://www.jstor.org/stable/26523180?seq=1&cid=pdf

Scott, J. C. (1998). Seeing like a state (Issue March). Yale University Press.

Senghor, K., Partelow, S., Herrera, G., & Osemwegie, I. (2023). Conflicting governance realities: Aligning historical and cultural practices with formal marine protected area co-management in Senegal. *Marine Policy*, *155*. https://doi.org/10.1016/j.marpol.2023.105706

van Laerhoven, F., Schoon, M., & Villamayor-Tomas, S. (2020). Celebrating the 30th anniversary of ostrom's governing the commons: Traditions and trends in the study of the commons, revisited. *International Journal of the Commons*, *14*(1), 208–224. https://doi.org/10.5334/ijc.1030

VeneKlasen, L., & Miller, V. (2007). Power and empowerment. In *A new weave of power, people and politics: The action guide for advocacy and citizen participation* (Vol. 6, Issue 2, pp. 131–134). Practical Action Publishing. https://doi.org/10.1016/0277-5395(83)90002-X

Walljasper, J. (2010). *All that we share: A field guide to the commons*. The New Press.

West, P., Igoe, J., & Brockington, D. (2006). Parks and peoples: The social impact of protected areas. In *Annual review of anthropology* (Vol. 35, pp. 251–277). https://doi.org/10.1146/annurev.anthro.35.081705.123308

Stefan Partelow is a professor at the Center for Life Ethics, University of Bonn. His research focuses on commons governance, collective action, and sustainability transformations in coastal and marine social–ecological systems. He combines empirical research and ethical reflection to promote just and adaptive approaches to environmental governance.

Open Access This chapter is licensed under the terms of the Creative Commons Attribution-NonCommercial-NoDerivatives 4.0 International License (http://creativecommons.org/licenses/by-nc-nd/4.0/), which permits any noncommercial use, sharing, distribution and reproduction in any medium or format, as long as you give appropriate credit to the original author(s) and the source, provide a link to the Creative Commons license and indicate if you modified the licensed material. You do not have permission under this license to share adapted material derived from this chapter or parts of it.

The images or other third party material in this chapter are included in the chapter's Creative Commons license, unless indicated otherwise in a credit line to the material. If material is not included in the chapter's Creative Commons license and your intended use is not permitted by statutory regulation or exceeds the permitted use, you will need to obtain permission directly from the copyright holder.

Justice and the Commons: A Review

Floriane Clement and Cécile Barnaud

Abstract When commons studies emerged as a new body of knowledge in the 1980s, environmental sustainability, rather than social justice, was their primary concern. This initial focus has influenced the theoretical and conceptual development of this field of study and the way commons scholars have discursively and practically engaged with issues of justice. Today, environmental justice has grown as a diverse and rich body of knowledge, and commons feature as a central issue in environmental justice movements and scholarship. Yet commons studies still rarely explicitly draw on 'justice' as an analytical concept or vocabulary of action. In this article, we analyse how justice has been conceptualised in both mainstream and critical commons studies, so as to explore how the fields of commons studies and environmental justice can mutually benefit from greater connections, both conceptually and empirically. Drawing on a historical narrative review, we start with examining how early research conducted under the Bloomington school of political economy has considered and conceptualised issues related to social justice, in relation to institutions. We then review the literature on commons that explicitly mentions justice or equity issues, drawing on both institutionalist studies of the commons, that stem from the Bloomington school, and critical studies on commons and commoning. We examine how these two strands have conceptualised (or not) justice, guided by the following questions: what are the main framings related to justice and commons and what is the implicit or explicit rationale for the use of justice in commons studies? How has justice been used analytically? Which forms of injustice have these studies highlighted and which ones have been under-addressed? What approaches to justice do they miss? In particular, how are gendered and intersectional forms of injustice addressed? Our findings highlight that institutionalist commons studies have largely focused on equity and distributive justice among community users but without paying much attention to how unequal distribution may be shaped by unequal power relationships and political-economic structures. This contrasts with the social and

F. Clement (✉) · C. Barnaud
Dynafor, INRAE, Toulouse University, Toulouse, France
e-mail: floriane.clement@inrae.fr

C. Barnaud
e-mail: cecile.barnaud@inrae.fr

© The Author(s) 2026

S. Partelow (ed.), *Ethics and the Commons*, Schriften zu Leben und Gesellschaft - Series on Life and Society 3, https://doi.org/10.1007/978-3-032-24246-4_2

environmental justice literature, which has relied on integrated framings of justice articulating distributive justice with issues of recognitional and procedural justice. Critical studies of the commons and commoning have broadened the dimensions and the subjects of justice considered, highlighting how collective action in the face of commons enclosure may be class-, ethnicity- and gender-based. We conclude with research perspectives to advance constructive engagements on justice within the commons literature, notably towards a reconsideration of nonhumans in commons studies.

The question of justice goes beyond existing property relations. It can be more about who needs something or who should have rights, than about who does have them.

Diane Rocheleau, 2006

1 Introduction

For the last two decades, the commons have become a rallying flag for global justice social movements (De Angelis & Harvie, 2013; Goldman, 1998; Martínez-Alier, 1997; Martinez-Alier et al. 2016; Schlosberg, 2004). On the one hand, these movements are rooted in people's lived experiences of extensive commons enclosures, such as land, sea and water grabs (Bennett et al., 2015; Dell'Angelo et al., 2017). On the other hand, at a more abstract level, the commons have become a central concept in alternative political agendas against neoliberal forms of injustice (Cumbers, 2015; Federici, 2018). In parallel, academic studies of environmental governance are increasingly relying on justice as a central analytical concept across different types of commons: forest (Forsyth & Sikor, 2013; He et al., 2021; Sikor, 2010; Suiseeya, 2017), water (Boelens et al., 2018, 2023; Ulloa, 2020; Zwarteveen & Boelens, 2014), biodiversity (Lenzi et al., 2023; Martin et al., 2016; Massarella et al., 2021; Sikor et al., 2014), oceans (Bennett et al., 2023; Blythe et al., 2023) and climate (Chatterton et al., 2013; Joshi, 2021; Newell et al., 2021; Okereke & Coventry, 2016; Schlosberg & Collins, 2014; Thomas & Twyman, 2005). This rapidly growing body of research on environmental justice advocates taking into account three interconnected dimensions of justice in relation to environmental changes and policies (Coolsaet, 2020; Schlosberg, 2007; Sikor et al., 2014), namely: distributive justice (who are the winners and losers of environmental changes and policies? who receives the benefits and who endures the costs, risks and efforts?), procedural justice (who participates in decision-making about environmental rules and policies?), and recognitional justice (what values, knowledge and worldviews are acknowledged and recognized in these rules and policies?) (Box 1).

Yet, strikingly, 'justice' is still rare in the work of commons scholars. A recent review of 50 years of published work on the study of the commons indicates that only 80 out of 4000 titles reviewed, i.e., 2% of the publications, mention 'environmental justice' (van Laerhoven et al., 2020), whereas some scholars defend that 'Ostrom's work contains no concept of justice' (Obeng-Odoom, 2016, 4). The lack of consideration of justice in early commons studies is not surprising, as commons scholarship initially emerged to identify the conditions that support the sustainable governance of common-pool resources, rather than to address justice and inequality-related challenges per se. The *raison d'être* of Elinor Ostrom's seminal book *Governing the Commons* (Ostrom, 1990), which inspired several generations of scholars around the world, was to demonstrate that community-based natural resource management did not necessarily lead to the over-exploitation and degradation of natural resources and could be at least as effective as privatisation or public management. Hence, the initial focus of commons studies on environmental sustainability issues may have lastingly influenced not only the development of the analytical and conceptual tools of this body of knowledge, i.e., studies drawing from the work of the Ostrom Workshop (cf. Barnett et al., 2020), but also the underlying values and discourses of the commons scientific community, who have overall poorly engaged with questions of inequality (Bousquet et al., 2015; Kashwan et al., 2021).

Social and environmental justice may refer to social movements, political discourses and theoretical concepts. **Social justice** refers in this chapter to justice among citizens in a society. **Environmental justice** refers to social justice issues related to environmental changes and policies. It mostly concerns justice among human beings, but some authors also include non-human species and ecosystems as subjects of justice.

There are plural conceptions and theories of social and environmental justice, which both developed from an initial concern with the fairness of the distribution of (environmental) goods and bads in society (**distributive justice**). This concern expanded to consider the underlying social, cultural and economic conditions responsible for unfair distribution. This leads to link distributive justice with **procedural justice**, related to participation in decision-making processes, and with**justice as recognition,** which refers to the recognition and respect of social and cultural differences, and, in particular, of the values, identities and knowledge systems of different social groups.

We consider**institutions** as the rules, norms and strategies that organise and regulate structured social interactions.

However, there is a need to scrutinize and further unpack how justice and commons have been empirically, normatively and conceptually connected within different bodies of scholarship working on commons, and how these connections have evolved over time. While commons scholarship is undoubtedly strongly rooted in Elinor Ostrom's heritage, it is neither fixed nor monolithic. The concept of commons has gained traction within multiple tangential fields, with different levels of connections with the seminal work of the Bloomington school, and this school is itself evolving and regenerating. Many scholars anchored in environmental justice or political ecology engage with the study of commons, but do not necessarily label their work as 'commons studies'. To understand how justice issues are addressed in commons scholarship, we need to embrace these diverse scholarly bodies of work.

We draw on a historical narrative literature review, i.e., an exploratory qualitative review, to investigate how justice has been considered in commons studies, and more largely how justice and commons have been articulated across multiples bodies of scholarship. Our objective is to identify trends and research gaps and to reflect on promising research avenues to further advance the consideration and conceptualisation of justice in commons studies.

In the first part of the chapter, we seek to roughly map commonalities and divergences in how social justice theories and institutional analysis conceptualise the relationships between justice, institutions and collective action. We first provide a brief overview on different theories of justice and how they conceive the role of institutions. We then examine the early foundational works of the Bloomington school of political economy, the research programme in institutional economics and political economy founded by Elinor and Vincent Ostrom, which has developed the institutionalist approach to the commons and expanded the concepts of polycentricity and self-governance. We analyse how Ostrom and her colleagues initially considered and conceptualised social and environmental justice. Following this analytical overview, we present in a second section the methods used for the literature review. Our analysis is guided by the following research questions:

What are the main framings related to justice and commons?
What is the implicit or explicit rationale for considering justice?
How has justice been used analytically?
Which forms of injustice have these studies highlighted?
Which ones have been under-addressed?

We conclude with the implications of our results for a fruitful dialogue between environmental justice and diverse strands of commons studies.

2 Justice and Institutions

2.1 *The Role of Institutions in Social and Environmental Justice Theories*

John Rawls' theory 'Justice as fairness' (1971) has been one of the most influential theories of social justice. It focuses on distributive justice, related to the distribution of goods in a society. John Rawls is considered as an institutionalist as his principles of justice apply to institutions rather than to individuals (Valentini, 2011). He postulates that in an ideal world, individuals would choose, under 'a veil of ignorance',[1] political principles of justice that can then be used to design a set of just institutions for

[1] Rawls posits that if individuals do not know their own identity or position in the society, they will not be subject to cognitive biases, related to their personal interests and desires, that would influence their choices of what are just institutions.

governing society (Rawls, 1971). In his view, developing fair basic institutions is a prerequisite for social justice.

In contrast, Amartya Sen criticizes Rawls' 'transcendental institutionalism' (2009) and proposes a comparative, rather than an ideal, approach of justice. Importantly, Sen challenges the assumption that just institutions will result in the right behaviours that will lead to a just society. Drawing on Sanskrit verses of ethics and jurisprudence, Sen makes a distinction between *niti* and *nyaya*, which both stand for 'justice' in Sanskrit. *Niti* refers to proper conducts and procedure while *nyaya* means judgement about what is observed in the real world. Sen describes Rawls' approach as *niti*-centered. On the contrary, Sen's 'idea of justice' defends the importance of *nyaya* and theorises justice as a deliberative process whereby different stakeholders are given the opportunity (and have the capability) to voice competing interests and values, deliberate on conflicting claims about what is an injustice in a specific context and reach acceptable decisions through reasoned arguments (Sen, 2009).

Rather than assuming compliant behaviour to a set of ideal institutions, Amartya Sen proposes to focus on people's actual freedoms and capabilities to live the life they choose. Amartya Sen and Marta Nussbaum considerably extend distributive justice theories by considering the capabilities that allow individuals to have both the opportunity and command over the goods distributed and to achieve the functionings that they value (Nussbaum, 2000; Nussbaum & Sen, 1993; Sen, 1999). For these scholars, a just distribution of goods cannot be intrisically considered as a just outcome, as long as the capabilities to turn goods into functionings are lacking.

Sen does not disregard the role that institutions can play towards greater social justice. However, rather than 'treating the institutions as themselves manifestations of justice' (Sen, 2009, 82), he regards institutions as a significant medium that facilitates the pursuit of justice by providing opportunities for public deliberation, i.e., by supporting freedom of speech, right to information and providing space for informed and interactive discussions. Hence, he values institutions not for their existence per se but for the extent to which they recognise and include diverse values, agendas and voices (ibid.). Justice '*is not simply the setting up of some specific [local] institutions, ... but the possibility and reach of public reasoning*' (ibid., xiii). He however does not specify what kind of institutions are required to reach public reasoning that would produce just outcomes, especially in the case of deep moral disagreement or entrenched social divisions along lines of caste or religion (Dryzek, 2013).

Critical and feminist theorists such as Iris Marion Young, and Nancy Fraser have defended the central role of recognition in social justice. Iris Marion Young (1990) posits that the institutionalised oppression and domination of particular social groups shape the distribution of material goods. Hence poor distribution and the non-recognition of social differences are intimately linked. In her view, the social and cultural recognition of marginalised groups needs to be supported by affirmative actions through institutions that foster inclusion and better representation in decision-making. Nancy Fraser similarly argues that distribution and (mis)recognition co-produce injustices and justices. She stresses the importance of political participation and deliberation while highlighting the existence of institutionalised barriers to equal participation in the public sphere (Fraser, 1990). She also defends that a globalising world has made even more critical to address not only the 'what' but also the 'who' of justice (Fraser, 2010). In her perspective, institutions do not only play a key role in enabling the creation of arenas where people can put forward their claims, but also in constituting subjects of justice: '*all those affected by a given social structure or institution have moral standing as subjects of justice in relation to it*' (Fraser, 2009, 24).

Environmental justice scholars have largely drawn on social justice theories but several environmental justice theorists (e.g., Schlosberg, 2007; Sikor et al., 2014) have also defended the need to ground our understanding of environmental justice based on the actual claims of social groups and citizens who experience forms of injustice (Coolsaet, 2020). As perceptions of both social and environmental injustice and related claims are plural (Schlosberg, 2007), how to acknowledge a plurality of justice claims brings forward the issues of participation and political representation. All forms of participation and representation are exclusionary, with a particular risk of concealing the voices and interests of marginalized social groups (Fraser, 1998; Young, 2002).

2.2 Theorization of Justice in Institutional Analysis and Commons Studies

A central challenge in commons scholarship has been one of collective action with an underlying concern for distributive justice, namely how to cooperate in ways that produce sufficient long-term benefits for all (Berge & van Laerhoven, 2011). Hence commons scholars have, from the outset, paid attention to issues of distributive justice, and have considered fairness or equity as a key outcome resulting from collective action. Early works of the Bloomington school of political economy have largely and explicitly drawn on Rawls' theory of social justice. Vincent and Elinor Ostrom share a similar belief and optimism in the key role institutions can play to achieving social justice (Ostrom, 1972). This is particularly prominent in early works of Vincent Ostrom on self-governing systems and polycentric governance (Ostrom, 1972, 1977). Vincent Ostrom et al. (1961) defined a polycentric system as a system "composed of (1) many autonomous units formally independent of one another, (2) choosing to act in ways that take account of others, (3) through processes of cooperation, competition, conflict, and conflict resolution" (Ostrom, 2014, p. 46). For the Ostroms, polycentricity was not just a form of governance or an institutional configuration. It was before all a political ideal that could lead to a more just society (Aligica & Tarko, 2012).

The Institutional Analysis and Development (IAD) framework is the foundational and central analytical framework developed and used by the Bloomington school. It is suited to the analysis of policy processes and has been extensively used to analyse institutional performance in commons management (Ostrom, 1999, 2005). We unpacked the framework to closely examine to which extent and how justice

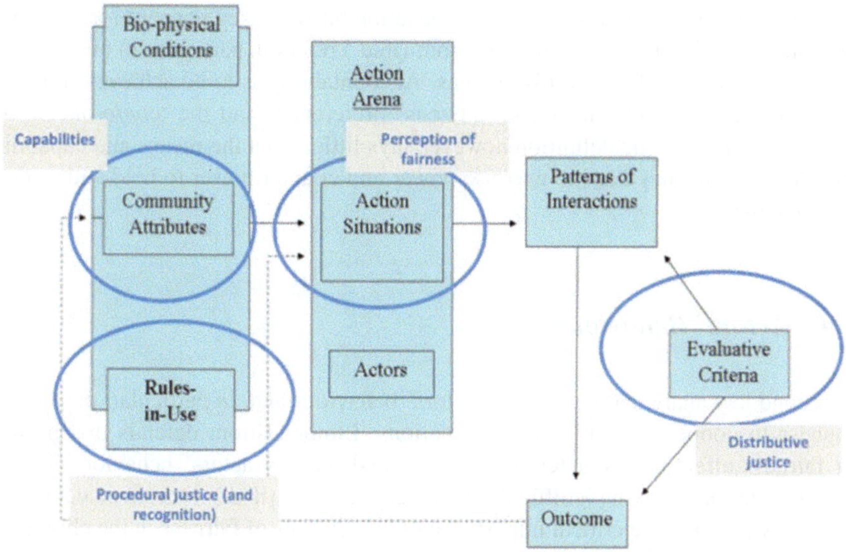

Fig. 1 Analytical entries used by commons scholars for considering justice in the IAD framework (adapted from Ostrom, 1999). The image depicts the IAD framework components, with circles around the most common entry points into justice in the framework used by scholars to examine justice issues

has been considered in different components of the framework and identified four main components (Fig. 1): the evaluative criteria to assess the outcomes of actors' interactions and decisions, the action situation, where actors interact with one another and make decisions, the community attributes and the rules-in-use, which affect the action arena.

2.3 *Evaluative Criteria and Distributive Justice*

The evaluative criteria of the outcomes are probably the most explicit integration of justice concerns into the IAD framework, notably with the inclusion of fairness (Ostrom et al., 1994) and equity (Ostrom, 1999) as evaluative criteria. Elinor Ostrom (ibid., 2011) distinguished two forms of equity: fiscal equivalence and re-distributional equity. Fiscal equivalence refers to the differential abilities of individuals who benefit from a service to bear the burden of contributing through their efforts to the service or to pay for it. Re-distributional equity refers to the need to provide services and facilities to 'benefit particularly needy groups' (1999, 34)—in line with Rawls' principle of difference (1972). Elinor Ostrom (ibid.) notes that the two forms of equity might be contradictory, as particularly needy groups might not be able to finance the services provided.

The evaluative criteria also include 'accountability', a key concept for recognition and procedural justice. Elinor Ostrom (ibid.) refers in particular to public officials being accountable towards citizens. Accountability is to be achieved through the aggregation of information on citizens' preferences and the '*conformance to general morality*'. The definition however says little about the means and capabilities that citizens may need to exercise power and counter power to hold authorities accountable.

2.4 Actors' Behaviour

A second key analytical entry is the actors' behaviour and, in particular, their willingness to cooperate within action situations. Elinor Ostrom defends that norms of fairness affect actors' internal valuation and thereby actors' behaviour within action situations, making it difficult to predict outcomes (ibid.). Ostrom et al. (1994) explicitly mention the critical importance of perceptions of fairness in the choice of rules that are considered appropriate. In turn, actors' perceptions of legitimacy and fairness affect their compliance to rules and their propensity to turn to free-riding behaviour (Ostrom, 2003). In recent work, Elinor Ostrom (2005, 263) goes a bit further by explicitly linking the perception of the fairness of distribution with procedural justice (participation in rules design), considering the extent to which rules meet local concepts of fairness. Her representation of human behaviour is more nuanced than that of John Rawls, who sees individuals as rational users who would mutually benefit from social cooperation (see Robeyns, 2008). On the contrary, Ostrom explicitly agrees with Sen that there are multiple models of rational behaviour:

> [W]hat has come to be called rational-choice theory in the social sciences is instead one model in a family of models useful when conducting formal analyses of human decisions in highly structured physical and institutional settings. As Amartya Sen points out in his presidential address to the American Economic Association, 'There are many different conceptions of rational behaviour of the individual'. The model of complete rationality most frequently used in rational-choice explanations is one model of these different conceptions. (Ostrom, 2003, 25 in Burbrige, 2016)

Elinor Ostrom moved away from the model of complete rational behaviour to draw on the concept of bounded rationality to consider actors' preferences and limited knowledge, and the influence of norms and strategies on human behaviour. However, her model of collective action heavily relies on individual human behaviour and pays little attention to social relations.

2.5 Heterogeneity of Capabilities

The Bloomington school has paid a great deal of attention to the issue of heterogeneity, which they initially articulated around three dimensions '*heterogeneity of capabilities, heterogeneity of preferences, and heterogeneity of beliefs and information*' (Aligica, 2014, 4). A rich body of knowledge has explored the effects of heterogeneity of capabilities (income, assets skills, techniques) (Andersson & Agrawal, 2011; Baland et al., 2007) and of social heterogeneity, looking at caste, gender and ethnicity (e.g. Agarwal, 2009; Baland & Platteau, 1996; Baland et al., 2007; Bardhan & Dayton-Johnson, 2002; Ruttan, 2006, 2008) as sub-components of the 'attributes of the community' (Fig. 1). However, the motivation in considering social differences lied in improving models of human behaviour and understandings of collective action and institutional performance (Aligica, 2014) rather than in supporting a better recognition of oppressed and marginalised social groups. The concept itself of 'heterogeneity' testifies of the apolitical stance adopted by commons scholars, who did not engage in acknowledging heterogeneity as social inequalities and power differences (Kashwan et al., 2021).

2.6 *Rules-In-Use*

Lastly, justice is considered in the rules-in-use and in the institutional design principles, which are to guide the design of the rules-in-use. The rules-in-use are decomposed into a set of seven rules. Three of them directly relate to procedural justice: the boundary rules, which define who can participate in the action situation and who is excluded; the position rules, which assign specific positions and authority to participating actors, thereby directly affecting their capabilities and responsibilities; and the information rules, which affect the kind of information available to participants. The latter however do not consider whether information might be unequally shared among participants. Although several traditional commons studies identify that boundary rules may include/exclude actors along social markers such as gender, class, caste and ethnicity (e.g. Ostrom, 1999), they usually assumed that the local community would have some objective discussion of what would be the best rules to follow and that all would have an equal voice—yet most often, social markers strongly dictate boundary rules themselves in ways that are not necessarily conscious and explicit (Cleaver, 2002).

Among the eight design principles proposed by Ostrom (1990), we found that only one principle includes a consideration for justice (more precisely, procedural justice): those affected by the rules can participate in their design (design principle 3).[2]

To sum-up, our review indicates that the concept of justice is not absent from the analytical tools developed and used by commons scholars. The way the Ostrom Workshop has considered issues of justice seems to closely align with Rawls' distributive

[2] One of the design principles (#7) refers to the recognition of users' rights to design their own institutions but as rightly pointed out by one reviewer, this cannot be considered as a form of recognition *as per* Fraser or other critical theorists's (e.g. Honneth) idea of recognition. They draw on Hegel's idea of freedom, which is based on the need for recognition and respect by others (Coolsaet & Néron 2020).

justice perspective. Yet the Ostroms have also strongly departed from a single institutional theory *a la* Rawls by adopting a pluralist approach closely engaging with institutional diversity (Agilica, 2014; Ostrom, 2005). E. Ostrom largely refers to Elster's work on local justice (see Ostrom, 2005). The latter criticizes Rawl's theory of justice and defends a place-based approach to understand 'why, at a particular time and place, a particular institution adopts a particular principle for allocating a particular good' (Elster, 1992, 135). In her late work, Ostrom also explicitly refers to the need for a pluralist approach to justice, proposed by Sen: '*Nor do we have a single normative theory of justice that can unambiguously be applied to all settings* (Sen, 2009)' (Ostrom, 2010a). However, the way justice has been integrated in the IAD framework is not really conducive to consider plural theories or plural perceptions of justice among actors. Notably, the action situation is conceptualised as a space where actors interact and negotiate, but has not been considered as a space where actors with different capabilities and voices debate to reach just decisions. Their initial objective was to explain actors' (non-)cooperative individual behaviour rather than to capture and understand ethical relations and goals.

The institutional design principles pay a limited attention to issues of procedural justice and self-recognition developed by social justice theorists, and do not consider issues of representation justice, as developed later in the justice literature, which addresses the question of who is entitled to make justice claims (Fraser, 2010; Thew et al., 2020). In general, the question of representation has been insufficiently problematized in the commons literature (McKay, 2023).

In addition, the IAD framework does not really support considering the processes and structures that produce inequalities and differential capabilities in the first place. For the last decade, several scholars who had engaged with mainstream commons studies have paid attention to inequalities and power relationships (Kashwan, 2016; Mudliar & Koontz, 2018, 2021), notably in the IAD framework (Brisbois et al., 2019; Clement, 2010; Kashwan et al., 2019) and in polycentric governance (Morrison et al., 2019), yet very few have explicitly drawn on environmental or social justice theories (e.g., Hayes & Murtinho, 2018).

Lastly, the overall framing which has guided the institutional analysis of the commons is not geared towards considering real-world justice issues. Whereas equity is one of the evaluative criteria, what is at stake is evaluating and improving the performance of commons management rather than redressing injustice. For instance, classical commons studies have generally not paid much attention to how commons are appropriated by the powerful at the expense of others (Boyce & Shelley, 2013).

After this overview of how the role of institutions has been conceptualised in social justice theories and how justice has been considered in institutional approaches to commons studies, we now interrogate through a narrative literature review (Hodgkinson & Ford, 2014; Sovacool et al., 2018) how the concept justice has been considered and used, both analytically and normatively, in both institutional and critical approaches to studies of the commons. In doing so, we ask:

> What are the main framings related to justice in commons studies?
> What is the implicit or explicit rationale for its use?

How has justice been used analytically?
Which forms of injustice have these studies highlighted?
Which ones have been under-addressed?

3 Methods

We combined some features of a systematic literature review, namely an explicit search approach based on the uniform application of inclusion and exclusion criteria, with a qualitative in-depth analysis of a sample of articles, which characterises a narrative review. The rules that we adopted to ensure transparency and reproducibility of the method are the following:

Scholarship using 'justice' as an analytical or normative concept and explicitly referring to the (environmental) 'commons': We only considered the articles that explicitly use the term 'justice' as an important analytical or normative concept and that made an explicit reference to 'commons', 'common-pool resources' or 'common-property resources'. In addition, we only considered environmental commons related to natural entities and elements of nature such as water, forest, atmosphere or wildlife, and excluded intangible and non-natural commons such as energy,[3] knowledge and digital commons, urban commons, etc. Our sample however includes environmental commons in urban spaces such as community gardens.

Types of studies: The review considered all types of studies, including theoretical papers on justice in commons management and governance, review articles, as well as empirical studies applying justice and related concepts for the study of common-pool resource management and governance in qualitative or quantitative case studies, comparative research, field studies, experiments, meta-analyses or modelling.

Types of publication: The review included journal articles and review articles. It excluded conference proceedings, books, book chapters and other items.

Language: The review only considered articles published in English, although we acknowledge that there is a large and vibrant body of literature on commons and justice in Latin America that might have mostly been published in Spanish.

It may seem arbitrary to draw a boundary between environmental justice articles that use the term 'commons' and those that consider natural resources without labelling these as 'commons'. Similarly, we could also have included research work on the commons that discuss equity, fairness, recognition or participation, irrespectively of the use of the term 'justice'. In our sample, several articles use 'justice' as a proxy for equity and fairness, but without referring to the environmental or social justice literature, and the line between these studies and those that focus on equity but do not use the concept of 'justice' is sometimes relatively thin. We acknowledge this limitation in our sample, which we try to overcome by drawing on research closely

[3] We excluded energy for two reasons: first, it was not part of the traditional commons and we are mostly interested in tracing how the traditional commons field has developed. Second, solar and wind energy are intangible and their features may thus be closer to those of knowledge commons rather than natural commons. There was no article focusing on oil or coal.

connected to the themes of justice and commons, but that was not identified in our search, to contextualise or discuss our results.

The search strategy started with an exhaustive search on the Web of Science™. We used the following research string: TOPIC: (Justice AND ("common-pool resource*" OR "common pool resource*" OR "common property" OR "the commons")). The search on the Web of Science returned 221 results, after filtering for selecting only journal articles and review articles. We assessed that the articles identified by the search met the inclusion criteria, based on their title, abstract, keywords and subject descriptors, and full text when necessary.

We also conducted a supplementary search on the Digital Library of the Commons, using the same research string, but which searched the full text of articles (there was no search option available to narrow the search in the abstract/title/keywords). This raised 778 results, we manually reviewed 200 of the most relevant results, which often implied scrolling the full text. This allowed assessing whether our main search was effective and whether the keywords used allowed retrieving most relevant articles, i.e. articles which explicitly articulated justice and the commons. Among the 200 results, we found only one article that met all our search criteria, i.e. Justice AND ("common-pool resource*" OR "common pool resource*" OR "common property" OR "the commons") in the title, abstract or keywords and 17 articles which did articulate justice and the commons, but did not mention both concepts in the title or abstract or keywords. It means we may have underestimated the relevant results by around 10% through our main search. Because of a lack of time, we could not conduct the manual search for the 578 remaining results but we considered the additional article that met all our search criteria. We hence reviewed 222 articles.

The references were stored and shared between the two co-authors through endnote libraries. A detailed reading of each paper allowed for a qualitative analysis, which was organised according to several variables and questions (Table 1).

The rationale for using justice questioned whether the authors' concerns were instrumental (e.g., conducting a more effective analysis), motivated by an ethical concern, or else. We then examined how justice was defined, if defined at all, and the social or environmental justice theories that were drawn upon. We notably examined the dimensions of justice (distributive justice, procedural justice, and recognition) and the community of justice (justice for whom?) considered in the paper. Finally, we questioned the conceptualization of human-nature relationships underlying the justice approach adopted in the paper, i.e. justice for humans in the face of environmental changes and environmental policies (as traditionally studied in environmental justice), justice for nature and non-human species in the era of Anthropocene (the focus of ecological justice and multispecies justice), or integration of both (recent development of environmental studies, also sometimes labelled 'socio-ecological justice'). The variable 'pluralism' refers to the recognition of a diversity of definitions of justice and worldviews (Martin et al., 2020).

For the qualitative analysis of a sample published before 2018, reviewer 1 reviewed first, followed by reviewer 2, independently. Discrepancies in reviewers' descriptions were resolved in meetings between the two reviewers. For the qualitative analysis of

Table 1 Variables and questions guiding the qualitative analysis of each paper in the review

Variable	Question
Rationale for using justice	What is the rationale for using justice in the paper?
Definition of justice	Is justice explicitly defined? How is it defined? From which bodies of literature does the definition draw from?
Pluralism	Are plural perceptions of justice acknowledged?
Analytical use	How is justice analytically used? e.g. as factor affecting the social-ecological system (e.g. actors' decisions), as criteria to evaluate the outcomes, as a guiding principle to design rules What are the implicit or explicit linkages between institutions and justice
Justice dimensions	Which dimensions of justice are considered in the paper? (distributive, procedural, recognition)
Community of justice	Justice for whom? At which scale/level is justice considered? Among which social groups?
Conceptualization of human-nature relations	Does the paper discuss forms of environmental or ecological/multispecies justice?

the sample published between 2018 and 2023, reviewer 1 and reviewer 2 analysed different articles independently.

Based on a preliminary analysis of the characteristics of our sample, we decided to consider two periods: one running until 2017 which gathers a large majority of articles that fall into an institutional approach of the commons, and another one from 2018 until 2023, which is characterised by a large increase in the number of publications and a larger share of articles that adopt a more critical approach to commons. We thus identified 2018 as a pivotal year in the sample and present our analysis disaggregated according to these two time periods. This temporal disaggregation allows better capturing the evolution in how commons and justice have been articulated, and analysing to which extent approaches drawing from environmental justice studies and political ecology have supported the development of new perspectives on justice and the commons.

4 Results

4.1 Selected Sample

We selected 69 journal articles that matched our inclusion criteria for an in-depth qualitative analysis. We can clearly observe a sharp increase in the number of publications engaging with both commons and justice since the mid-2010s (Fig. 2): two thirds of the articles we selected were published in the last six years of the 25-year period considered.

Beyond this rise in the total number of publications, we also observed an increasing number of articles published in the last six years that we would position in the broad fields of environmental justice/critical commons studies/political ecology.

If we take a reflexive stance and look at our sample considering issues of epistemic justice, we note that the first authors of reviewed articles are mostly male (58%)—white males represent 41% of the sample. Less than 15% of the first authors work in academic organisations based in Asia, Africa and Latin America, which also reflects our sample bias related to the focus on English language publications. Research sites are relatively equally shared between Europe and North America on the one hand, and Asia, Africa and Latin America on the other hand.

Land represents 25% of the resources studied over the whole sample and regarding the two periods. It is not surprising considering enduring and new forms of inequality in land access and property rights, and the recent surges in land grabbing that have profoundly affected property relations across the world (Borras & Franco, 2012; Borras et al., 2011; Dell'Angelo et al., 2017; Razavi, 2003; Peters, 2004). In the 1996–2017 sample, the second most studied commons are fisheries, which is aligned with its historical importance in institutional commons studies (van Laerhoven et al., 2020). Yet it rarely comes up in the second time period –on the contrary, many studies examine 'commons' in general, e.g., social movements related to commons

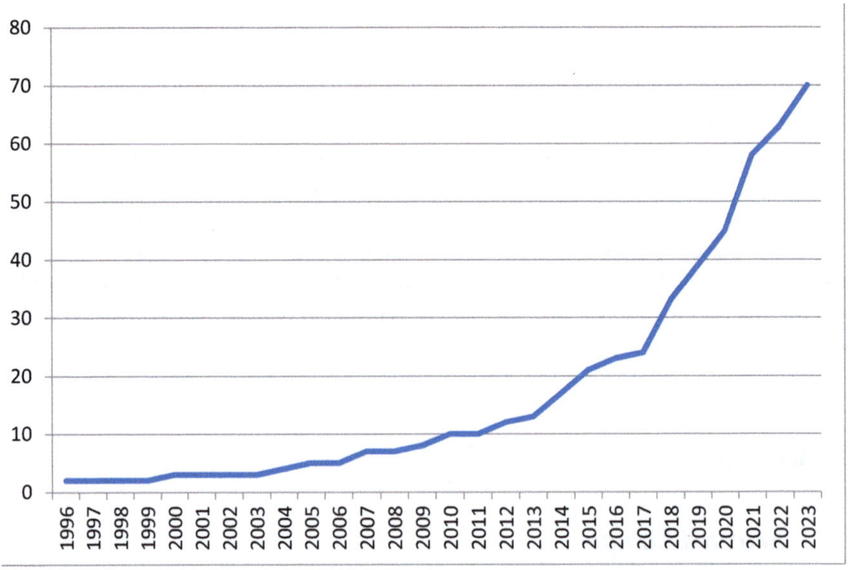

Fig. 2 Historical trend in the number of publications meeting our selection criteria in 1995–2020. The image depicts a bar graph with an upward trendline (linear from 1995–2013; exponential from 2013–2023) showing the number of publications per year and over time of literature relating to justice and the commons

enclosure or commons and commoning as governance systems, without necessarily focusing on a specific type of common.

Climate issues are totally missing in the first period and relatively under-addressed in the second period, despite a growing body of scholarship on climate justice. It could mean that the climate justice scholarship has in general not conceptualised climate or the atmosphere as a global or planetary common (Ostrom, 2010b; Rockström et al., 2024) (with a few exceptions (Joshi, 2021)).

4.2 Framings and Articulations of Justice and Commons: Grasping Justice Through an Institutional Approach

The reviewed articles display a wide range of engagement with the concept of environmental and social justice, as well as with the theoretical literature on justice. Overall, few articles define justice. We noted that a large number of the articles published before 2019 equate justice with equity or fairness, or use these concepts interchangeably without defining them. More than a third of these articles do not engage with any literature on social or environmental justice. We identified two main types of articulations between commons and justice during this early period.

Firstly, most of the selected articles published during this period draw on the concept of distributive justice to evaluate the effects of community- or project-led common resources management on the distribution of costs and benefits, mostly within a community or among resource users, but without considering how unequal distribution may stem from power relationships and social identities such as class, gender and ethnicity. The subjects of justice are undifferentiated resource users within a community. In a similar vein, but adding a conceptual question and an attention to gender-based discrimination and marginalisation, Choi (2009) assesses whether the economic efficiency of fishery management at the individual level compromises or is aligned with social justice at the collective level. These articles rely on empirical data, based on case studies. Whereas these analyses 'add' distributive justice into their analysis of the outcomes of environmentally sustainable commons management, Paavola (2007) goes further by challenging, at a conceptual and analytical level, the dominant focus of studies of institutional performance on efficiency. Paavola starts from the new institutional approach, but he defends that institutional choice is a matter of social justice, rather than a matter of efficiency. What is more, he does not stop at issues of distributive justice, but, drawing on Fraser (2001) and Schlosberg (1999), he proposes to integrate procedural and recognition justice issues into an extended analytical framework for environmental governance, which acknowledges a pluralism of values. Yet he does not reflect about how justice and plural values may be shaped by gender, ethnicity, or class. Yet, his article stands out as a pioneering work bridging an institutional approach on the commons and critical theories on social and environmental justice.

Secondly, a number of the selected articles published in this period draw on the concept of justice to improve the analysis of the factors affecting the functional performance of institutions. DeCaro et al. (2015) analyse how perceptions of justice influence actors' behaviours and, notably, their willingness to cooperate. This research question continues to drive several studies on commons and justice published more recently. For instance, Running et al. (2019) examine whether local water users' perceptions of fairness and efficacy affects the social acceptance of a water policy reform in the USA, and He et al. (2020) defend the need to analyse local people's perceptions of justice to understand why they resisted forest privatisation initiated through the Chinese Collective Forest tenure reform. Weesie (2019) proposes to take into account local perceptions of justice in order to improve Ostrom's design principles (1990). He et al. (2021) and Luo and He (2023) offer a more critical stance as they argue that local perceptions of justice are a more important factor than the adoption of rules in accordance with the design principles, based on evidence from community-based forestry and community-led irrigation water management in China.

Most of the reviewed studies published during the first period adopt an institutional approach to consider justice in commons management and governance and we can easily map these early perspectives onto the IAD framework (cf. Figure 1 earlier). A large majority of this work analyses outcomes in terms of distributive justice/equity, aligning with earlier work considering equity as an evaluative criterion. Another relatively large number of studies focuses on the action situation and shows that perceptions of fairness influence individual behaviour and collective action, sharing with the work of Ostrom and her colleagues the objective to debunk the so-called 'tragedy of the commons' by challenging the assumption that individuals are rational egoists.

We noted that many sub-variables of the IAD framework that could support a better consideration of justice in institutional analysis, identified in section II, have not been used or have been under-used. For instance, few studies look at how integrating justice in institutional design. Issues related to the design principles such as the recognition of the community's rights to develop their own rules, or government's accountability towards citizens are barely considered. There was only one study that used a capability approach to assess the effect of environmental conservation programmes from a human development perspective (Bockstael & Berkes, 2017).

To conclude on this first period, our results evidence that the rationale for considering justice has focused on improving analyses of institutional performance, by capturing a broader range of outcomes or of factors affecting actors' compliance or resistance to rules. It is noteworthy that only a small share of the reviewed articles consider justice because it is an issue of interest for the communities located in their case study (e.g. Chatterton et al., 2013). In contrast, this has been largely developed in the second period we considered, as we shall see in the next sub-section.

4.3 *New Directions for Environmental Justice and Commons Scholarship*

In the 2019–2023 period, a growing body of scholarship with a political ecology/critical perspective offers original ways to articulate justice and commons, providing new directions for commons studies.

First, a large number of studies in this second period have considered environmental justice as a claim and/or a driver for collective action to defend commons and commons access (Dell'Angelo et al., 2021; Fanari, 2022; Kidd, 2020; Villamayor-Tomas & García-López, 2018). The rationale to consider justice departs from the previous period, where the objective was to improve the analysis of institutional performance. In this case, it stems from empirical observations, from the lived reality of local people organising themselves to protect commons and 'their ways of living and surviving' (Hernández Vidal et al., 2023, 372). Originally, environmental justice has been closely tied to social movements, as 'the vocabulary and repertoire of EJ action was born with the Civil Rights Movement' in the USA (Hernández Vidal et al., 2023, 370). Similarly, there has been a long history of rural and urban 'commons movements' around the world with explicit claims to defend the commons (Villamayor-Tomas et al., 2023). Dell'Angelo et al. (2021) note that there is a clear linkage between just and environmentally sustainable commons management, as commons grabbing is often associated with unsustainable exploitation of natural resources. Scholars interested in the intersection between social movements and commons have unravelled how such forms of collective action help to protect and recognise communities' rights and capacity to manage commons (Fanari, 2022; Villamayor-Tomas & García-López, 2018, 2021). However, social movements led in the name of 'the commons' often largely go beyond what some observers would label an 'environmental struggle'. They may be rooted in historical processes of colonialism, authoritarian regimes and enduring forms of oppression of politically weak groups (Fanari, 2022; Hernández Vidal et al., 2023; McKay, 2023).

Interestingly, Fanari (2022) evidences how indigenous people's claims for forest justice in India are supported by a legal institutional basis, the Forest Rights Act, 2006, that grants forest resources rights to indigenous people. How legal rules and the formal recognition of communities' rights to manage their resources by the State can support collective action for greater social and environmental justice has been underexplored in earlier studies on the commons. However, rules alone are not sufficient to secure access (Ribot & Peluso, 2003) and local resistance is critical for a real political transformation. This opens new research perspectives on how institutions can support critical consciousness as citizens are able to acknowledge structures of oppression and imagine that they have the right to bring change. For Apostolopoulou et al. (2022), it is specific places that shape people's political subjectivities by providing them with "frameworks for interpreting whether injustices have been done and whether collective and contentious responses are merited" (Nicholls et al., 2016, 4). Whereas social resistance emerges in particular places for historically and geographically specific reasons, it is also connected to broader inequalities/unequal or undemocratic structures (Sultana, 2023). Apostolopoulou et al. (2022) thus defend the need to cross both spatial and temporal scales in order to create enduring grassroots innovations that can gain wider support and solidarity. Whereas local communities may lead, in many cases, relatively invisible forms of day-to-day resistance to defend commons (Apostolopoulou et al., 2022; Fanari, 2022), social mobilisation is strongly associated with risks of dispossession, violence and coercion (Dell'Angelo et al., 2021), especially directed towards women or feminised subjects, such as children and the elderly (Hernández Vidal et al., 2023). It may hence be necessary for local communities to form alliances with local and international actors, not only to increase the potential for radical political transformation but also to protect their lives. The choice of alliances is critical as these raise issues of power and epistemic justice as international non-governmental organisations (INGOS) may convey and impose western framings of environmental justice and conservation (Martin et al., 2016). A second important new perspective on justice and commons envisions commons and commoning as a political alternative to achieve greater social and environmental justice. In this body of work, following Silke Helfrich and David Bollier (2015) and Pierre Dardot and Christian Laval (2015), commons are defined in opposition to capitalist systems (see McPhearson et al., 2021; Rutt, 2020) and inseparably intertwined with commoning, i.e. the social practices and social relations that make and remake the commons (Clement et al., 2019). This perspective may also consider the role of social movements but with a focus on how their political agenda gears towards building new forms of commoning, rather than on how they resist against dispossession or exclusion. In this view, commons and collective management tend to be promoted as intrinsically more just than other types of governance. Some scholars argue that commons and commoning promote 'citizen-led innovation and participation' (McPhearson et al., 2021, 8), an argument that resonates with Sen's idea of justice, according to whom institutions are just as they promote the recognition of plural voices and values. Others defend that commoning offers not only alternative spaces to resist but also to transform capitalist and unequal dominant power structures (Dobbin, 2021). This argument aligns with the views of political philosophers Iris Marion Young and Nancy Fraser, as it broadens the subjects of justice beyond resource users to encompass all social groups affected by marginalisation and oppression.

Some scholars question the intrinsically more just, democratic and empowering nature of the commons (see McKay, 2023). For instance, Maurits de Jongh (2021) interrogates the linkage between justice and the commons, which have been heralded as a political alternative to both state and market appropriation and 'a general principle of society's reorganization' (Dardot & Laval, 2015, 143). He challenges the anti-statist arguments by arguing that commons can also turn into sites of injustice. Hence, if commons are to be a just and sustainable alternative to privatisation, financiarisation and commodification, they must be embedded within a strong public system that counter these injustices. Rebecca Rutt (2023) reaches similar conclusions when examining the extent to which commoning can develop through urban farming in Copenhagen, among an increasingly neoliberalised city and state. She points to the role that public policies, notably geared towards degrowth, can play in supporting more sustainable, just and convivial community initiatives like urban gardens. Although her article was not in our reviewed sample, Andrea Nightingale (2019) also cautions against idealised notions of commons and commoning, and evidences how they are not exempt from power relationships but also create forms of inclusion and exclusion. Many articles from this second period acknowledge the gendered nature of environmental injustice (see Hernández Vidal et al., 2023) and how gendered inequalities may be a driver for women-led mobilisation. Yet very few do adopt or advocate for an intersectional analysis or reflect on how gendered maldistribution and misrecognition are intertwined, as noted by Fraser (2007).

Also noteworthy is the large predominance of anthropocentric approaches in our sample. As seen above, the conceptual articulations between commons and justice are diverse and evolving, but in all these approaches, the communities of justice remain human, focusing on intra-human justice issues. They have indeed a common purpose of highlighting and defending the rights and claims of local communities, or subgroups within these communities, in the face of threats such as capitalism and subsequent environmental degradation, but also unfair environmental policies. Only two articles in our sample explicitly mention issues of justice for nature and non-human species (Wirawan et al., 2023; Yaka, 2019). Building an original and underexplored bridge between justice and commons, Özge Yaka (ibid.) looks at human struggles for environmental commons through the concept of socio-ecological justice, which extends the community of justice to non-human by framing the relational existence of human and non-human ecologies as a matter of justice itself.

Lastly, interpretive perspectives articulate justice and commons as interlinked discourses and claims. Justice may be a catalyser for the emergence of collective action, commoning practices and new solidarities (Chatterton et al., 2013; Clark, 2019), or a rallying flag that legitimates social movements (Hönig, 2014; Upton, 2014), especially when local communities face expert or technocratic discourses (Dobbin, 2021) or discourses supported by coalitions of powerful international actors. Rights-based discourses can transform institutions as political weapons that largely go beyond their role of legal tools (Clark, 2019; Upton, 2014). Claiming a 'commons custodian' identity may help to secure a voice in global environmental justice arenas (Upton, 2014). Issues of recognition are here closely linked to procedural justice, i.e., who participates in decision-making (Hernández Vidal et al., 2023). Viviana Asara (2020) also shows in the case of the Indignados movement in Barcelona how claiming rights to the commons may not just involve standing up against commons privatisation but may also express an appeal for a just society for humans and nonhumans with a different way of relating to each other. In this case, discourses become voices of resistance to capitalism that both open up and help to reproduce new radical imaginaries.

5 Conclusion: Towards Fruitful Dialogues Between Environmental Justice and Commons

In 2006, when the International Association for the Studies of Common Property (IASCP) dropped the 'P' of property to become the International Association for the Studies of the Commons (IASC), Diane Rocheleau, then Executive Council Member of the IASC, outlined in a note, published in the quarterly publication of the IASC, that this change in the name of the Association would support a better consideration of justice in commons studies (Rocheleau, 2006).

This narrative review has indeed evidenced a dramatic increase in how commons have been articulated with justice issues since the mid-2010s, but our analysis suggests that the literature drawing from critical studies has played an important role in renewing and reinvigorating how justice has been considered by engaging a dialogue with commons studies. Yet it is likely that this dialogue may not have taken place with the 'studies of common property' which were a priori narrowing the analytical focus to the political, legal and economic aspects of property relations, neglecting the moral, social and ecological dimensions of the commons, as argued by Rocheleau (ibid.).

A first trend of research, prevailing until the late 2010s, has integrated justice following an institutional approach to commons studies (with some exceptions), with two different areas of investigation: a first area relates to issues of distributive justice and aims at capturing the costs and benefits of commons management among community users, and; a second area relates to how perceptions of fairness affect actors' behaviour and their compliance to rules. Such work helps to develop some of the 'seeds of justice' that were already in the IAD framework, but not all have been explored, e.g. a better consideration of procedural justice into institutional design or communities' and of individuals' capabilities to achieve the functionings they value through commons management.

The second trend, which particularly developed from the late 2010s onwards, has more largely drawn on critical approaches and does not necessarily fit into existing institutional analysis frameworks. Critical geography, political ecology and environmental justice scholarship broaden the analytical lenses of commons scholarship in several ways.

Firstly, this trend includes the study of new research objects, notably environmental justice movements, relatively understudied in the commons scholarship until recently. Such forms of resistance have taken place for decades, although they might not have been visible to Westerners' eyes (Lohmann, 1995), but they have significantly increased in the face of rising commons grabbing and enclosure (Dell'Angelo et al., 2021). In the face of contemporary forms of extraction and dispossession, commons-related challenges have shifted from their sustainable management to their defence and/or collective forms of re-appropriation. Whereas Ostrom and her colleagues shared a concern for communities' exclusion from commons access and management for conservation purpose, they did not examine the moral aspect of this exclusion or its social impacts, but rather attempted to show that it was not necessarily an efficient way to conserve commons. This positioning may have been considered as a more effective entry point to influence conservation actors and policy-makers, who, following Hardin, may also defend that 'injustice is preferable to total ruin' (Hardin, 1968, 1247). The studies of social movements highlight the risks of criminalisation of actors and increased/new forms of leadership or the recognition of rights as potential outcomes of collective action, thus broadening the type of outcomes initially considered by Ostrom. However, the analysis of power relations and inequalities in environmental justice movements remains underdeveloped in our sample (except Kidd, 2020), despite a few exceptions on gendered participation in water struggles in Latin America and South Asia (Joshi, 2014; Laurie, 2011).

Secondly, environmental justice scholarship requires us to pay attention to the roots of distributive (in)justice, namely: why is there 'mal-distribution' and how does it sustain, to simultaneously consider recognition and distributive justice *a la* Fraser. Considering claims for recognition might support greater intersectional approaches to understand inequalities, i.e., how gender, ethnicity, class, etc. intersect to produce injustices—we see a slightly greater attention to gender in this second trend, although it remains quite marginal in the analysis. It also brings attention to plural perceptions of justice: what does 'fairness' or benefit' mean for different people (Forsyth & Sikor, 2013) and hence pushes for re-framing the action situation as an arena for debate and not just as a place where actors participate, communicate and make choices, as envisioned in the IAD framework. This strand also relies on original methods based on collaborative knowledge production (Hernández Vidal et al., 2023), with an underlying ambition to produce emancipatory knowledge, rather than knowledge that reproduces and sustains social inequalities (Sullivan, 2000).

Lastly, critical approaches on commons and commoning open up new discourses and imaginaries that allow conceptualising new political and social alternatives to capitalist and neoliberal economies and analysing grassroots innovations that attempt at experimenting such radical ideas on the ground. It can offer a new framing and a new vocabulary for commons scholarship as proposed by Velicu and García-López: 'From this perspective, sustaining the earth's commons is not a mere technical management of resources (in space) but a struggle to perform common liveable relations (in time)' (2018, 3).

Conversely, institutional approaches to commons studies also have something to offer the environmental justice and commoning literature. First, studies on polycentric governance can support a nuanced understanding of how local forms of commoning may link up with complex forms of governance at higher scales (Ostrom, 2010b) and articulate with the state (see Mansbridge, 2014). In addition, institutional approaches enable a fine analysis of the rules and norms governing collective action in environmental justice movements and commoning processes to understand who is allowed to participate, how decisions are taken or who is represented. Furthermore, as Yaka (2019) argues, a socio-ecological approach can advance current theories of justice by acknowledging the role that biophysical attributes may play in how distributive, procedural and recognition issues emerge. Commons and social-ecological systems (SES) studies can provide useful insights on how to integrate ecological variables into the analysis. Scholars who have a good knowledge of institutional approaches to the commons and an interest in justice have played and will continue to play a significant role in bridging in these bodies of scholarship and in deepening the conversation that is just beginning.

A yet underexplored perspective of this dialogue is the consideration of justice for nature and nonhumans. By justice for nature, we mean considering the distribution of damages across human and ecosystems, recognising the existence, characteristics and integrity of ecosystems, defining social and political institutions that recognise the natural world in decision-making procedures and avoiding the disruption of ecosystems' capabilities and functioning (Schlosberg, 2007). Both commons studies and environmental justice literature are initially anchored in anthropocentric approaches, centred on the interests and voices of diverse human groups. Although environmental justice generally aims to reconcile social justice and environmental concerns, it focuses mainly on how diverse human groups are affected and their voice taken into account, be it in environmental changes, struggles and policies. In the same way, although commons studies were initially concerned with environmental sustainability, they are largely based on anthropocentric and utilitarian representations of nature, considering the interests of resource users as the main goal of governance arrangements (Lara, 2015). It is not surprising since one of their *raison d'être* was to move away from the ecocentric approaches that prevailed in biodiversity conservation in 1960s (Mace, 2014). Based on the idea of preserving nature for itself, and from human threats, these ecocentric approaches failed to recognize that human communities had the capacities and interest of preserving natural resources (Mathevet et al., 2018a, 2018b). After several decades of development of anthropocentric approaches in environmental governance, there is a revival of more ecocentric approaches, and attempts to reconcile both (Shastri, 2013; Washington et al., 2021). The idea of considering nature and nonhumans as subjects of justice, as proposed by ecological justice scholars (Low & Gleeson, 1998; Wienhues, 2020), has been feeding increasingly vivid debates among environmental justice scholars (Cooke, 2020; Schlosberg, 2014). Also controversial is the idea of attributing rights to nature (Chapron et al., 2019), and the increasing number of experiences where nature elements are allocated legal personhood and their rights defended in court (O'Donnell, 2020; Wesche, 2021). Some scholars argue that nonhumans cannot have rights, in the sense of social contracts and enforceable claims, and remain dependant on humans, who have their own interests and interplays (Bétaille, 2019). Yet a number of environmental scientists advocate for extended communities of justice, including all living beings, highlighting the common destiny of humans and nonhumans, the continuity of their relationship, human's ability to empathize for nonhuman, and humans' knowledge on living conditions for nonhumans (Baxter, 2005; Cooke et al., 2020; Schlosberg, 2007; Yaka, 2019). This ecological turn in environmental justice echoes an emerging interest for more-than-humans in critical commons and commoning studies (Bresnihan, 2015; Yaka, 2019), and more broadly other non-utilitarian environmental philosophies, such as ethics of care, environmental stewardship or ecological solidarity (Bennett et al., 2018; Bieling et al., 2020; Mathevet et al. 2018a, 2018b; Welchman, 1999). A dialogue between commons, justice and other scholars would fruitfully address these debates and the vast number of questions they raise, which are admittedly controversial, but surely urgent and necessary to address in the era of the Anthropocene.

Acknowledgements Earlier versions of this book chapter benefited from feedback from participants in the biennal IASC conference 2019 in Lima and in the webinar of the EJJE (Environmental Justice/Justice Environnementale) research network held on 31/03/2022. We thank Bruno Bouet for reviewing and commenting an earlier version.

Authors' Contributions: Conceptualization, methodology: F.C. (lead) & C.B. (supporting). Investigation & formal analysis: F.C. (lead) & C.B. (supporting); Writing—original draft: F.C.; Writing—review and editing: F.C. & C.B.

References

Agarwal, B. (2009). Gender and forest conservation: The impact of women's participation in community forest governance. *Ecological Economics, 68*(11), 2785–2799. https://doi.org/10.1016/j.ecolecon.2009.04.025

Aligica, P. D. (2014). *Institutional diversity and political economy. The Ostroms and beyond*. Oxford University Press.

Aligica, P. D., & Tarko, V. (2012). Polycentricity: From Polanyi to Ostrom, and beyond. *Governance, 25*(2), 237–262. https://doi.org/10.1016/j.gloenvcha.2011.03.004

Andersson, K., & Agrawal, A. (2011). Inequalities, institutions, and forest commons. *Global Environmental Change, 21*(3), 866–875. https://doi.org/10.1016/j.gloenvcha.2011.03.004

Apostolopoulou, E., Bormpoudakis, D., Chatzipavlidis, A., Cortés Vázquez, J. J., Florea, I., Gearey, M., & Wahby, N. (2022). Radical social innovations and the spatialities of grassroots activism: Navigating pathways for tackling inequality and reinventing the commons. *Journal of Political Ecology, 29*(1), 144–188. https://doi.org/10.2458/jpe.2292

Asara, V. (2020). Untangling the radical imaginaries of the Indignados' movement: Commons, autonomy and ecologism. *Environmental Politics*, 1–25. https://doi.org/10.1080/09644016.2020.1773176

Baland, J.-M., & Platteau, J.-P. (1996). *Halting degradation of natural resources: Is there a role for rural communities*? Food and Agriculture Organisation (FAO).

Baland, J. M., Bardhan, P., & Bowles, S. (Eds.). (2007). *Inequality, cooperation, and environmental sustainability*. Princeton University Press.

Bardhan, P., & Dayton-Johnson, J. (2002). Unequal irrigators: Heterogeneity and commons management in large-scale multivariate research. In E. Ostrom, T. Dietz, N. Dolšak, P. C. Stern, S. Stonich, E. U. Weber, & Committee on the Human Dimensions of Global Change, Division of Behavioral and Social Sciences and Education (Eds.), *The drama of the commons* (pp. 86–112). National Academy Press.

Barnett, A. J., Partelow, S., Frey, U., García-Lozano, A., del Mar Mancha-Cisneros, M., Oberlack, C., & Whitney, C. K. (2020). Defining success in the commons. *International Journal of the Commons, 14*(1), 366–387. https://doi.org/10.5334/ijc.994

Baxter, B. (2005). *A theory of ecological justice*. Routledge.

Bennett, N. J., Alava, J. J., Ferguson, C. E., Blythe, J., Morgera, E., Boyd, D., & Côté, I. M. (2023). Environmental (in)justice in the Anthropocene ocean. *Marine Policy, 147*, Article 105383. https://doi.org/10.1016/j.marpol.2022.105383

Bennett, N. J., Govan, H., & Satterfield, T. (2015). Ocean grabbing. *Marine Policy, 57*, 61–68. https://doi.org/10.1016/j.marpol.2015.03.026

Bennett, N. J., Whitty, T. S., Finkbeiner, E., Pittman, J., Bassett, H., Gelcich, S., & Allison, E. H. (2018). Environmental stewardship: A conceptual review and analytical framework. *Environmental Management, 61*, 597–614. https://doi.org/10.1007/s00267-017-0993-2

Berge, E., & van Laerhoven, F. (2011). Governing the Commons for two Decades: A complex story. *International Journal of the Commons, 5*(2), 160–187. https://doi.org/10.18352/ijc.325

Bétaille, J. (2019). Rights of nature: Why it might not save the entire world. *Journal for European Environmental & Planning Law, 16*(1), 35–64. https://doi.org/10.1163/18760104-01601004

Bieling, C., Eser, U., & Plieninger, T. (2020). Towards a better understanding of values in sustainability transformations: Ethical perspectives on landscape stewardship. *Ecosystems and People, 16*(1), 188–196. https://doi.org/10.1080/26395916.2020.1786165

Blythe, J. L., Gill, D. A., Claudet, J., Bennett, N. J., Gurney, G. G., Baggio, J. A., & Zafra-Calvo, N. (2023). Blue justice: A review of emerging scholarship and resistance movements. *Cambridge Prisms: Coastal Futures, 1*, Article e15. https://doi.org/10.1017/cft.2023.4

Bockstael, E., & Berkes, F. (2017). Using the capability approach to analyze contemporary environmental governance challenges in coastal Brazil. *International Journal of the Commons, 11*(2), 799–822. https://doi.org/10.18352/ijc.756

Boelens, R., Escobar, A., Bakker, K., Hommes, L., Swyngedouw, E., Hogenboom, B., & Wantzen, K. M. (2023). Riverhood: Political ecologies of socionature commoning and translocal struggles for water justice. *The Journal of Peasant Studies, 50*(3), 1125–1156. https://doi.org/10.1080/03066150.2022.2120810

Boelens, R., Perreault, T., & Vos, J. (2018). *Water justice*. Cambridge University Press.

Borras Jr, S. M., & Franco, J. C. (2012). Global land grabbing and trajectories of agrarian change: A preliminary analysis. *Journal of Agrarian Change, 12*(1), 34–59. https://doi.org/10.1111/j.1471-0366.2011.00339.x

Borras, S. M., Jr., Hall, R., Scoones, I., White, B., & Wolford, W. (2011). Towards a better understanding of global land grabbing: An editorial introduction. *The Journal of Peasant Studies, 38*(2), 209–216. https://doi.org/10.1080/03066150.2011.559005

Bousquet, F., Robbins, P., Peloquin, C., & Bonato, O. (2015). The PISA grammar decodes diverse human–environment approaches. *Global Environmental Change, 34*, 159–171. https://doi.org/10.1016/j.gloenvcha.2015.06.0

Boyce, J. K., & Shelley, B. G. (Eds.). (2013). *Natural assets. Democratizing environmental ownership*. Island Press.

Bresnihan, P. (2015). The more-than-human commons: From commons to commoning. In S. Kirwan, L. Dawney, & J. Brigstocke (Eds.), *Space, power and the commons* (pp. 93–112). Routledge.

Brisbois, M. C., Morris, M., & de Loë, R. (2019). Augmenting the IAD framework to reveal power in collaborative governance—An illustrative application to resource industry dominated processes. *World Development, 120*, 159–168. https://doi.org/10.1016/j.worlddev.2018.02.017

Chapron, G., Epstein, Y., & López-Bao, J. V. (2019). A rights revolution for nature. *Science, 363*(6434), 1392–1393. https://doi.org/10.1126/science.aav560

Chatterton, P., Featherstone, D., & Routledge, P. (2013). Articulating climate justice in Copenhagen: Antagonism, the commons, and solidarity. *Antipode, 45*(3), 602–620. https://doi.org/10.1111/j.1467-8330.2012.01025.x

Choi. (2009). Eochongye social capital in Jujeon-dong, Ulsan. *Korea Journal, 49*(1), 5–32.

Clark, C. (2019). Water justice struggles as a process of commoning. *Community Development Journal, 54*(1), 80–99. https://doi.org/10.1093/cdj/bsy052

Cleaver, F. (2002). Reinventing institutions: Bricolage and the social embeddedness of natural resource management. *The European Journal of Development Research, 14*(2), 11–30. https://doi.org/10.1080/714000425

Clement, F. (2010). Analysing decentralised natural resource governance: Proposition for a "politicised" institutional analysis and development framework. *Policy Sciences, 43*(2), 129–156. https://doi.org/10.1007/s11077-009-9100-8

Clement, F., Harcourt, W. J., Joshi, D., & Sato, C. (2019). Feminist political ecologies of the commons and commoning. *International Journal of the Commons, 13*(1), 1–15. https://doi.org/10.18352/ijc.972

Cooke, S. (2020). Justice beyond humanity. In B. Coolsaet (Ed.), *Environmental justice* (pp. 279–290). Routledge.

Cooke, B., Landau-Ward, A., & Rickards, L. (2020). Urban greening, property and more-than-human commoning. *Australian Geographer, 51*(2), 169–188. https://doi.org/10.1080/00049182.2019.1655828

Coolsaet, B. (2020). *Environmental justice: Key issues.* Routledge.

Coolsaet, B., & Néron, P. Y. (2020). Recognition and environmental justice. In B. Coolsaet (Ed.), *Environmental justice* (pp. 52–63). Routledge.

Cumbers, A. (2015). Constructing a global commons in, against and beyond the State. *Space and Polity, 19*(1), 62–75. https://doi.org/10.1080/13562576.2014.995465

Dardot, P., & Laval, C. (2015). Commun. Essai sur la révolution du XXIe siècle. La Découverte.

De Angelis, M., & Harvie, D. (2013). The commons. In M. Parker, G. Cheney, V. Fournier, & C. Land (Eds.), *The Routledge companion to alternative organization* (pp. 280–294). Routledge.

de Jongh, M. (2021). Public goods and the commons: Opposites or complements? *Political Theory, 49*(5), 774–800. https://doi.org/10.1177/00905917209799

DeCaro, D. A., Janssen, M. A., & Lee, A. (2015). Synergistic effects of voting and enforcement on internalized motivation to cooperate in a resource dilemma. *Judgment and Decision Making, 10*(6), 511–537. https://doi.org/10.1017/S1930297500006975

Dell'Angelo, J., D'Odorico, P., Rulli, M. C., & Marchand, P. (2017). The tragedy of the grabbed commons: Coercion and dispossession in the global land rush. *World Development, 525*(92), 1–12. https://doi.org/10.1016/j.worlddev.2016.11.005

Dell'Angelo, J., Navas, G., Witteman, M., D'Alisa, G., Scheidel, A., & Temper, L. (2021). Commons grabbing and agribusiness: Violence, resistance and social mobilization. *Ecological Economics, 184*, Article 107004. https://doi.org/10.1016/j.ecolecon.2021.107004

Dobbin, K. B. (2021). Environmental justice organizing as commoning practice in groundwater reform: Linking movement and management in the quest for more just and sustainable rural futures. *Elementa: Science of the Anthropocene, 9*(1), 00173. https://doi.org/10.1525/elementa.2020.00173

Dryzek, J. S. (2013). The deliberative democrat's Idea of Justice. *European Journal of Political Theory, 12*(4), 329–346. https://doi.org/10.1177/1474885112466784

Elster, J. (1992). *Local justice: How institutions allocate scarce goods and necessary burdens.* Russell Sage Foundation.

Fanari, E. (2022). Struggles for just conservation: An analysis of India's biodiversity conservation conflicts. *Journal of Political Ecology, 28*(1), 1–31. https://doi.org/10.2458/jpe.5214

Federici, S. (2018). *Re-enchanting the world: Feminism and the politics of the commons.* PM Press/Kairow

Forsyth, T., & Sikor, T. (2013). Forests, development and the globalisation of justice. *The Geographical Journal, 179*(2), 114–121. https://doi.org/10.1111/geoj.12006

Fraser, N. (1990). Rethinking the public sphere: A contribution to the critique of actually existing democracy. *Social Text, 25*(26), 56–80. https://doi.org/10.2307/466240

Fraser, N. (1998). *Social justice in the age of identity politics: Redistribution, recognition.* Sage.

Fraser, N. (2001). Recognition without ethics? *Theory, Culture and Society, 18*(2–3), 21–42. https://doi.org/10.1177/02632760122051760

Fraser, N. (2007). Feminist politics in the age of recognition: A two-dimensional approach to gender justice. *Studies in Social Justice, 1*(1), 23–35. https://doi.org/10.26522/ssj.v1i1.979

Fraser, N. (2009). *Scales of justice: Reimagining political space in a globalizing world.* Columbia University Press.

Fraser, N. (2010). Who counts? Dilemmas of justice in a postwestphalian world. *Antipode, 41*(s1), 281–297. https://doi.org/10.1111/j.1467-8330.2009.00726.x

Fraser, N., & Honneth, A. (2003). *Redistribution or Recognition?: A political-philosophical exchange*. Verso.
Goldman, M. (Ed.). (1998). *Privatizing nature: Political struggles for the global commons*. Pluto Press.
Hardin, G. (1968). The tragedy of the commons. *Science, 162*(3859), 1243–1248. https://doi.org/10.1126/science.162.3859.1243
Hayes, T., & Murtinho, F. (2018). Communal governance, equity and payment for ecosystem services. *Land Use Policy, 79*, 123–136. https://doi.org/10.1016/j.landusepol.2018.08.001
He, J., Kebede, B., Martin, A., & Gross-Camp, N. (2020). Privatization or communalization: A multi-level analysis of changes in forest property regimes in China. *Ecological Economics, 174*, Article 106629. https://doi.org/10.1016/j.ecolecon.2020.106629
He, J., Martin, A., Lang, R., & Gross-Camp, N. (2021). Explaining success on community forestry through a lens of environmental justice: Local justice norms and practices in China. *World Development, 142*, Article 105450. https://doi.org/10.1016/j.worlddev.2021.105450
Helfrich, S., & Bollier, D. (2015). Commons. In G. D'Alisa, F. Demaria, & G. Kallis (Eds.), *Degrowth: A vocabulary for a new era*. Routledge.
Hernández Vidal, N., Merlinsky, G., & Bolados, P. (2023). Defending the commons: New frontiers in Latin American perspectives on environmental justice. *Sociological Inquiry, 93*(2), 370–391. https://doi.org/10.1111/soin.12525
Hodgkinson, G. P., & Ford, J. K. (2014). Narrative, meta-analytic, and systematic reviews: What are the differences and why do they matter? *Journal of Organizational Behaviour, 35*(S1), S1–S5. https://doi.org/10.1002/job.1918
Honneth, A. (1992). Integrity and disrespect: Principles of morality based on the theory of recognition. *Political Theory, 20*(2), 187–201. https://doi.org/10.1177/0090591792020002001
Hönig, P. (2014). Civil society and land use policy in Uganda: The Mabira forest case. *Africa Spectrum, 49*(2), 53–77. https://doi.org/10.1177/000203971404900203
Joshi, D. (2014). Feminist solidarity? Women's engagement in politics and the implications for water management in the Darjeeling Himalaya. *Mountain Research and Development, 34*(3), 243–254. https://doi.org/10.1659/MRD-JOURNAL-D-13-00097.1
Joshi, S. (2021). *Climate change justice and global resource commons: Local and global postcolonial political ecologies*. Routledge. https://doi.org/10.4324/9780429346231
Kashwan, P. (2016). Integrating power in institutional analysis: A micro-foundation perspective. *Journal of Theoretical Politics, 28*(1), 5–26. https://doi.org/10.1177/0951629815586877
Kashwan, P., MacLean, L. M., & García-López, G. A. (2019). Rethinking power and institutions in the shadows of neoliberalism: (An introduction to a special issue of world development). *World Development, 120*, 133–146. https://doi.org/10.1016/j.worlddev.2018.05.026
Kashwan, P., Mudaliar, P., Foster, S. R., & Clement, F. (2021). Reimagining and governing the commons in an unequal world: A critical engagement. *Current Research in Environmental Sustainability, 3*, Article 100102. https://doi.org/10.1016/j.crsust.2021.100102
Kidd, D. (2020). Standing rock and the Indigenous commons. *Popular Communication, 18*(3), 233–247. https://doi.org/10.1080/15405702.2020.1781862
Lara, A. (2015). Rationality and complexity in the work of Elinor Ostrom. *International Journal of the Commons*, *9*(2), 573–594. https://doi.org/10.18352/ijc.468
Laurie, N. (2011). Gender water networks: Femininity and masculinity in water politics in Bolivia. *International Journal of Urban and Regional Research, 35*(1), 172–188. https://doi.org/10.1111/j.1468-2427.2010.00962.x
Lenzi, D., Balvanera, P., Arias-Arévalo, P., Eser, U., Guibrunet, L., Martin, A., & Pascual, U. (2023). Justice, sustainability, and the diverse values of nature: Why they matter for biodiversity conservation. *Current Opinion in Environmental Sustainability*, *64*, 101353. https://doi.org/10.1016/j.cosust.2023.101353
Lohmann, L. (1995). Visitors to the commons. In B. R. Taylor (Ed.), *Ecological resistance movements: The global emergence of radical and popular environmentalism* (pp. 109–126). SUNY Press.

Low, N., & Gleeson, B. (1998). *Justice, society and nature.* Routledge.
Luo, D., & He, J. (2023). Just commons. *International Journal of the Commons, 17*(1), 141–154. https://doi.org/10.5334/ijc.1203
Mace, G. M. (2014). Whose conservation? *Science, 345*(6204), 1558–1560. https://doi.org/10.1126/science.125470
Mansbridge, J. (2014). The role of the state in governing the commons. *Environmental Science & Policy, 36*, 8–10. https://doi.org/10.1016/j.envsci.2013.07.006
Martin, A., Coolsaet, B., Corbera, E., Dawson, N. M., Fraser, J. A., Lehmann, I., & Rodriguez, I. (2016). Justice and conservation: The need to incorporate recognition. *Biological Conservation, 197*, 254–261. https://doi.org/10.1016/j.biocon.2016.03.021
Martin, A., Armijos, M. T., Coolsaet, B., Dawson, N., AS Edwards, G., Few, R., & White, C. S. (2020). Environmental justice and transformations to sustainability. *Environment: Science and Policy for Sustainable Development, 62*(6), 19–30. https://doi.org/10.1080/00139157.2020.1820294
Martínez-Alier, J. (1997). Environmental justice (local and global). *Capitalism Nature Socialism, 8*(1), 91–107. https://doi.org/10.1080/10455759709358725
Martinez-Alier, J., Temper, L., Del Bene, D., & Scheidel, A. (2016). Is there a global environmental justice movement? *The Journal of Peasant Studies, 43*(3), 731–755. https://doi.org/10.1080/03066150.2016.1141198
Massarella, K., Nygren, A., Fletcher, R., Büscher, B., Kiwango, W. A., Komi, S., & Percequillo, A. R. (2021). Transformation beyond conservation: How critical social science can contribute to a radical new agenda in biodiversity conservation. *Current Opinion in Environmental Sustainability, 49*, 79–87. https://doi.org/10.1016/j.cosust.2021.03.005
Mathevet, R., Bousquet, F., Larrère, C., & Larrère, R. (2018a). Environmental stewardship and ecological solidarity: Rethinking social-ecological interdependency and responsibility. *Journal of Agricultural and Environmental Ethics, 31*, 605–623. https://doi.org/10.1007/s10806-018-9749-0
Mathevet, R., Bousquet, F., & Raymond, C. M. (2018). The concept of stewardship in sustainability science and conservation biology. *Biological Conservation, 217*, 363–370. https://doi.org/10.1016/j.biocon.2017.10.015
McKay, S. (2023). Common-pool resources and democracy. *Critical Review of International Social and Political Philosophy*, 1–24. https://doi.org/10.1080/13698230.2023.2185382
McPhearson, T., Raymond, C. M., Gulsrud, N., Albert, C., Coles, N., Fagerholm, N., & Vierikko, K. (2021). Radical changes are needed for transformations to a good Anthropocene. *Npj urban sustainability, 1*(1), 5. https://doi.org/10.1038/s42949-021-00017-x
Morrison, T. H., Adger, W. N., Brown, K., Lemos, M. C., Huitema, D., Phelps, J., & Hughes, T. P. (2019). The black box of power in polycentric environmental governance. *Global Environmental Change, 57*, Article 101934. https://doi.org/10.1016/j.gloenvcha.2019.101934
Mudliar P., & Koontz, T. M. (2018). The muting and unmuting of caste across inter-linked action arenas: inequality and collective action in a community-based watershed group. *International Journal of the Commons, 12*(1), 225–248. https://doi.org/10.18352/ijc.807
Mudliar, P., & Koontz, T. M. (2021). Locating power in Ostrom's design principles: Watershed management in India and the United States. *Society & Natural Resources, 34*(5), 639–658. https://doi.org/10.1080/08941920.2020.1864535
Newell, P., Srivastava, S., Naess, L. O., Torres Contreras, G. A., & Price, R. (2021). Toward transformative climate justice: An emerging research agenda. *Wiley Interdisciplinary Reviews: Climate Change, 12*(6), Article e733. https://doi.org/10.1002/wcc.733
Nicholls, W., Miller, B., & Beaumont, J. (Eds.). (2016). *Spaces of contention: Spatialities and social movements.* Routledge.
Nightingale, A. J. (2019). Commoning for inclusion? Political communities, commons, exclusion, property and socio-natural becomings. *International Journal of the Commons, 13*(1), 16–35. https://doi.org/10.18352/ijc.927

Nussbaum, M. C. (2000). *Women and human development: The capabilities approach*. Oxford University Press.
Nussbaum, M. C., & Sen, A. K. (1993). *The quality of life*. Clarendon Press.
Obeng-Odoom, F. (2016). The meaning, prospects, and future of the commons: Revisiting the legacies of Elinor Ostrom and Henry George. *American Journal of Economics and Sociology, 75*(2), 372–414. https://doi.org/10.1111/ajes.12144
O'Donnell, E. (2020). Rivers as living beings: Rights in law, but no rights to water? *Griffith Law Review, 29*(4), 643–668. https://doi.org/10.1080/10383441.2020.1881304
Okereke, C., & Coventry, P. (2016). Climate justice and the International Regime: Before, during, and after Paris. *Wiley Interdisciplinary Reviews: Climate Change, 7*, 834–851. https://doi.org/10.1002/wcc.419
Ostrom, E. (1990). *Governing the commons. The evolution of institutions for collective action*. Cambridge University Press.
Ostrom, E. (1999). Institutional rational choice. An assessment of the institutional analysis and development framework. In P. A. Sabatier (Ed.), *Theories of the policy process* (pp. 35–71). Westview Press.
Ostrom, E. (2003). Toward a behavioral theory linking trust, reciprocity, and reputation. In E. Ostrom & J. Walker (Eds.), *Trust and reciprocity: Interdisciplinary lessons from experimental research* (pp. 19–79). Russell Sage Foundation.
Ostrom, E. (2005). *Understanding institutional diversity*. Princeton University Press.
Ostrom, E. (2011). Background on the institutional analysis and development framework. *Policy Studies Journal, 39*, 7–27. https://doi.org/10.1111/j.1541-0072.2010.00394.x
Ostrom, E. (2010a). Beyond markets and states: Polycentric governance of complex economic systems. *American Economic Review, 100*(3), 641–672. https://doi.org/10.1257/aer.100.3.641
Ostrom, E. (2010b). Polycentric systems for coping with collective action and global environmental change. *Global Environmental Change, 20*, 550–557. https://doi.org/10.1016/j.gloenvcha.2010.07.004
Ostrom, E., Gardner, R., & Walker, J. (1994). *Rules, games, and common-pool resources*. University of Michigan Press.
Ostrom, V. (1972). Polycentricity. In *Annual meeting of the American Political Science Association*. The American Political Science Association.
Ostrom, V. (1977). Some problems in doing political theory: A response to Golembiewski's "critique." *American Political Science Review, 71*(4), 1508–1525. https://doi.org/10.2307/1961492
Ostrom, V. (2014). Polycentricity: The structural basis of self-governing systems. In Ostrom E. & Ostrom V. (Eds), *Choice, rules and collective action: The Ostrom's on the study of institutions and governance* (pp. 45—60). ECPR Press.
Ostrom, V., Tiebout, C. M., & Warren, R. (1961). The organization of government in metropolitan areas: A theoretical inquiry. *American Political Science Review, 55*(4), 831–842.
Paavola, J. (2007). Institutions and environmental governance: A reconceptualization. *Ecological Economics, 63*(1), 93–103. https://doi.org/10.1016/j.ecolecon.2006.09.026
Rawls, J. (1971). *A theory of justice*. Oxford University Press.
Razavi, S. (2003). Introduction: Agrarian change, gender and land rights. *Journal of Agrarian Change, 3*(1–2), 2–32. https://doi.org/10.1111/1471-0366.00049
Peters, P. E. (2004). Inequality and social conflict over land in Africa. *Journal of Agrarian Change, 4*(3), 269–314. https://doi.org/10.1111/j.1471-0366.2004.00080.x
Ribot, J. C., & Peluso, N. L. (2003). A theory of access. *Rural sociology, 68*(2), 153–181. https://doi.org/10.1111/j.1549-0831.2003.tb00133.x
Robeyns, I. (2008). Ideal theory in theory and practice. *Social Theory and Practice, 34*(3), 341–362. https://doi.org/10.5840/soctheorpract200834321
Rocheleau, D. (2006). Putting property in context: From common property to the properties of the commons. *The Commons Digest, 2*, 6–7.

Rockström, J., Kotzé, L., Milutinović, S., Biermann, F., Brovkin, V., Donges, J., & Steffen, W. (2024). The planetary commons: A new paradigm for safeguarding Earth-regulating systems in the Anthropocene. *PNAS, 121*(5), Article e2301531121. https://doi.org/10.1073/pnas.2301531121

Running, K., Burnham, M., & Du Bray, M. V. (2019). Perceptions of fairness in common-pool resource access: Farmer responses to new agricultural water use restrictions in Idaho. *Environmental Sociology, 5*(4), 405–415. https://doi.org/10.1080/23251042.2019.1643548

Rutt, R. L. (2020). Cultivating urban conviviality: Urban farming in the shadows of Copenhagen's neoliberalisms. *Journal of Political Ecology, 27*(1), 612–634.

Ruttan, L. M. (2006). Sociocultural heterogeneity and the commons. *Current Anthropology, 47*(5), 843–853.

Ruttan, L. M. (2008). Economic heterogeneity and the commons: Effects on collective action and collective goods provisioning. *World Development, 36*(5), 969–985. https://doi.org/10.1016/j.worlddev.2007.05.005

Schlosberg, D. (1999). *Environmental Justice and the New Pluralism: The Challenge of Difference for Environmentalism.* Oxford University Press.

Schlosberg, D. (2004). Reconceiving environmental justice: Global movements and political theories. *Environmental Politics, 13*(3), 517–540. https://doi.org/10.1080/0964401042000229025

Schlosberg, D. (2007). *Defining environmental justice theories movements and nature.* Oxford University Press.

Schlosberg, D. (2014). Ecological justice for the anthropocene. In M. Wissenburg & D. Schlosberg (Eds.), *Political animals and animal politics* (pp. 75–89). Palgrave Macmillan.

Schlosberg, D., & Collins, L. B. (2014). From environmental to climate justice: Climate change and the discourse of environmental justice. *Wiley Interdisciplinary Reviews: Climate Change, 5*, 359–374. https://doi.org/10.1002/wcc.275

Sen, A. (2009). *The Idea of Justice.* Harvard University Press.

Sen, A. K. (1999). *Development as freedom.* Anchor.

Shastri, S. C. (2013). Environmental ethics anthropocentric to eco-centric approach: A paradigm shift. *Journal of the Indian Law Institute, 55*(4), 522–530.

Sikor, T. (2010). Forest justice: Towards a new agenda for research and practice? *Journal of Integrative Environmental Sciences, 7*(4), 245–250. https://doi.org/10.1080/1943815X.2010.530878

Sikor, T., Martin, A., Fisher, J., & He, J. (2014). Towards an empirical analysis of justice in ecosystem governance. *Conservation Letters, 7*(6), 524–532. https://doi.org/10.1111/conl.12142

Sovacool, B. K., Axsen, J., & Sorrell, S. (2018). Promoting novelty, rigor, and style in energy social science: Towards codes of practice for appropriate methods and research design. *Energy Research & Social Science, 45*, 12–42. https://doi.org/10.1016/j.erss.2018.07.007

Suiseeya, K. R. M. (2017). Contesting justice in global forest governance: The promises and pitfalls of REDD+. *Conservation and Society, 15*(2), 189–200. https://doi.org/10.4103/cs.cs_15_104

Sullivan, S. (2000). Getting the science right, or introducing science in the first place? Local 'facts', global discourse—'Desertification' in north-west Namibia. In P. Stott & S. Sullivan (Eds.), *Political ecology: Science, myth and power* (pp. 15–44). Edward Arnold.

Sultana, F. (2023). Whose growth in whose planetary boundaries? Decolonising planetary justice in the Anthropocene. *Geo: Geography and Environment, 10*(2), e00128. https://doi.org/10.1002/geo2.128

Thew, H., Middlemiss, L., & Paavola, J. (2020). "Youth is not a political position": Exploring justice claims-making in the UN climate change negotiations. *Global Environmental Change, 61*, Article 102036. https://doi.org/10.1016/j.gloenvcha.2020.102036

Thomas, D. S. G., & Twyman, C. (2005). Equity and Justice in climate change adaptation amongst natural-resource-dependent societies. *Global Environmental Change, 15*(2), 115–124. https://doi.org/10.1016/j.gloenvcha.2004.10.001

Ulloa, A. (2020). The rights of the Wayúu people and water in the context of mining in La Guajira, Colombia: Demands of relational water justice. *Human Geography, 13*(1), 6–15. https://doi.org/10.1177/1942778620910894

Upton, C. (2014). The new politics of pastoralism: Identity, justice and global activism. *Geoforum, 54*, 207–216. https://doi.org/10.1016/j.geoforum.2013.11.011

Valentini, L. (2011). A paradigm shift in theorizing about justice? A critique of Sen. *Economics & Philosophy, 27*(3), 297–315. https://doi.org/10.1017/S0266267111000228

van Laerhoven, F., Schoon, M., & Villamayor-Tomas, S. (2020). Celebrating the 30th Anniversary of Ostrom's *Governing the Commons*: Traditions and trends in the study of the commons, revisited. *International Journal of the Commons, 14*(1), 208–224. https://doi.org/10.5334/ijc.1030

Velicu, I., & García-López, G. (2018). Thinking the commons through Ostrom and Butler: Boundedness and vulnerability. *Theory, Culture & Society, 35*(6), 55–73. https://doi.org/10.1177/026 32764187573

Villamayor-Tomas, S., & García-López, G. (2018). Social movements as key actors in governing the commons: Evidence from community-based resource management cases across the world. *Global Environmental Change, 53*, 114–126. https://doi.org/10.1016/j.gloenvcha.2018.09.005

Villamayor-Tomas, S., & García-López, G. A. (2021). Commons movements: Old and new trends in rural and urban contexts. *Annual Review of Environment and Resources, 46*, 511–543. https://doi.org/10.1146/annurev-environ-012220-102307

Villamayor-Tomas, S., García-López, G., & D'alisa, G. (2023). Commons regimes at the crossroads: Environmental justice movements and commoning. In S. Villamayor-Tomas & R. Muradian (Eds.), *The Barcelona school of ecological economics and political ecology: A companion in honour of Joan Martinez-Alier* (pp. 219–233). Springer International Publishing.

Washington, H., Piccolo, J. J., Gomez-Baggethun, E., Kopnina, H., & Alberro, H. (2021). The trouble with anthropocentric hubris, with examples from conservation. *Conservation, 1*(4), 285–298. https://doi.org/10.3390/conservation1040022

Weesie, R. (2019). Towards adaptive commons: A case study of Agro-Pastoral dams in northern Ghana. *Sustainability, 11*(2), 319. https://doi.org/10.3390/su11020319

Welchman, J. (1999). The virtues of stewardship. *Environmental Ethics, 21*(4), 411–423. https://doi.org/10.5840/enviroethics19992146

Wesche, P. (2021). Rights of nature in practice: A case study on the impacts of the Colombian Atrato river decision. *Journal of Environmental Law, 33*(3), 531–555. https://doi.org/10.1093/jel/eqab021

Wienhues, A. (2020). *Ecological justice and the extinction crisis: Giving living beings their due.* Bristol University Press.

Wirawan, R., McIntyre-Mills, J. J., Makaulule, M., Lethole, P. V., Pitsoane, E., Arko-Achemfuor, A., & Romm, N. R. (2023). Together we can grow: Resourcing the commons through pathways to wellbeing. *Systemic Practice and Action Research, 36*(5), 641–690. https://doi.org/10.1007/s11213-022-09613-z

Yaka, Ö. (2019). Rethinking justice: Struggles for environmental commons and the notion of socio-ecological justice. *Antipode, 51*(1), 353–372. https://doi.org/10.1111/anti.12422

Young, I. M. (1990). *Justice and the politics of difference.* Princeton University Press.

Young, I. M. (2002). *Inclusion and democracy.* Oxford University Press.

Zwarteveen, M. Z., & Boelens, R. (2014). Defining, researching and struggling for water justice: Some conceptual building blocks for research and action. *Water International, 39*(2), 143–158. https://doi.org/10.1080/02508060.2014.891168

Floriane Clement is a senior researcher at the Dynafor research laboratory and works for the National Research Institute for Agriculture, Food and the Environment (INRAE) in Toulouse, France. Her work examines the implementation of environmental policies, drawing on (feminist) political ecology and commons studies.

Cécile Barnaud is a senior researcher at the Dynafor laboratory, INRAE, Toulouse University, France. Her interdisciplinary work bridges environmental justice, participatory action research, and agroecological transitions in rural landscapes. She leads the French section of the European "Just Scapes" project which explores fair transformations in land use under climate policy pressures.

Open Access This chapter is licensed under the terms of the Creative Commons Attribution-NonCommercial-NoDerivatives 4.0 International License (http://creativecommons.org/licenses/by-nc-nd/4.0/), which permits any noncommercial use, sharing, distribution and reproduction in any medium or format, as long as you give appropriate credit to the original author(s) and the source, provide a link to the Creative Commons license and indicate if you modified the licensed material. You do not have permission under this license to share adapted material derived from this chapter or parts of it.

The images or other third party material in this chapter are included in the chapter's Creative Commons license, unless indicated otherwise in a credit line to the material. If material is not included in the chapter's Creative Commons license and your intended use is not permitted by statutory regulation or exceeds the permitted use, you will need to obtain permission directly from the copyright holder.

Many Worlds, One Earth: Towards a Diverse Ethics of the Commons

Erin Rizzato Devlin

Abstract In a time of ecological crisis, the notion of the commons can become a means to promote social justice, equity and sustainability responses to the challenges of a changing landscape. This chapter aims to reflect on the urgent need to rethink and practice the commons by understanding the ontological, epistemic and planetary backdrop within which this concept is embedded. Before considering the ethical aspects of governing the commons globally, an introductory discussion on cosmology and plural ontologies will take place to understand the role of pluralism in the context of globalisation. These onto-epistemic considerations aim to shed light on the implications of dichotomies such as North/South and global/local, particularly in relation to power and relationships to the commons. Although most struggles against unequal power are local, the question of how we can politically surpass a single struggle can only be achieved by imagining an accumulation of localisms (Massey, 2015). The power of the global is precisely that it differs from place to place, whilst building solidarity amongst diversity. After having sketched a philosophical case for moving beyond these dichotomies, practical policy directions will be suggested in the hope of shining some light on how an appropriate conceptualisation of the commons can guide ethical governance amongst diversity. The legal recognition of the rights of nature, or future generation acts are, amongst others, proposals that incorporate this pluralistic approach and radically shift the conceptualisation of the commons for the future.

E. Rizzato Devlin (✉)
Equitable Energy Research CIC, Lerwick, Scotland
e-mail: erin@equitable.energy

Center for Energy Ethics, University of St Andrews, St Andrews, Scotland

© The Author(s) 2026

S. Partelow (ed.), *Ethics and the Commons*, Schriften zu Leben und Gesellschaft - Series on Life and Society 3, https://doi.org/10.1007/978-3-032-24246-4_3

"Our Worlds are at war"

Aílton Krenak, indigenous writer[1]

1 Introduction

The commons is becoming an increasingly popular notion due to its versatility and suitability in response to planetary challenges. The thought underlying this discussion is that taking a commons approach to natural resources will not only be necessary when facing the anthropogenic depletion of natural resources, but also beneficial to transforming the dominant economic model which has contributed to the processes currently threatening the Earth system (Chapin et al., 2011; Järvensivu et al., 2018). A commons can be generally defined as a social system where resources are shared by a collective of users/producers who are responsible for defining how these resources are used, produced, distributed and circulated through horizontal forms of governance (De Angelis & Harvie, 2014). Today, however, this common ownership coexists alongside neoliberal notions of economic growth, which has often replaced the commons approach led by local communities with private ownership or state-controlled goods, therefore privileging a form of enclosure or co-optation over the collective or community-based management of resources. By reconceptualising the relationship between people and public resources, this common ownership can stimulate new ways to imagine how property, ownership and hierarchies may be reconfigured to preserve resources and make their access and distribution more equitable.

However, when exporting this model to the planetary scale to collectively govern the commons across cultures and boundaries, particularly in response to the challenges of an ever-connected globe, the practical problem of global collective action is coupled with a more fundamental ethical question regarding differing worlds, which has often been ignored. In fact, challenges arise around different ways of enacting, creating, (re)producing the world and its resources, as well as the knowledges, relationships and values that accompany them. As the contemporary world is now confronted with wicked problems such as globalisation, climate change and the energy crisis, a renewed imaginative effort must be made to bring those plural worlds into one while preserving their diversities, their contextualities, their identities. This effort must therefore take into account that all beings see (or "represent") the world in the same way, though what changes is the world they see: what a human being sees as blood, a jaguar may see as maize beer, thus "see[ing] in the *same* way as we do *different* things [or 'worlds']" (Viveiros de Castro, 1998, p. 478). For this reason, the notion of the commons can become a useful instrument to bridge and understand how natural resources are shared amongst multiple human cosmologies and sketch an

[1] Originally published in the December 2019 issue of the Brazilian magazine *Olympio: Literatura e Arte*. The testimony was related orally to José Eduardo Gonçalves and Maurício Meirelles, and then transcribed.

ethical guideline for interacting with a shared earth in a way that embraces pluralism yet does not avoid, but rather contains, the potential for conflict.

2 Our Worlds: A Preliminary Cosmology

The idea that different "worlds" exist has been used for a long time to divide various countries into categories. Since the Cold War, the term "First World", as opposed to "Second" or "Third", has been largely employed to draw nations from the capitalist Western block away from the Eastern communist block beyond the Iron Curtain, as well as from non-aligned or neutral countries. Though widely used, it was never clear whether the "Three Worlds" discourse adhered to precise criteria, or was simply a vague notion to cluster together a diverse collection of countries to the advantage of the West, as opposed to the "Rest" (Tomlinson, 2003). Other collective descriptions such as "developing world", "less-developed world" and the "South" later emerged, along with "Third World", to describe countries from Africa, Asia, the Middle East, Latin America and the Pacific Islands.

The reason why the expression "First World" is disturbing is due to the fact that this is not simply a metaphor. Rather the concept of a "First World", as opposed to a "Third World" destined to emulate or replicate the former's processes of development, has grown out of specific forms of global interactions which have seen European nation states rise into power through imperialism and global markets. This reminds us that the subjugated South "is a result of our collective lack of imagination, our inability in our present difficult circumstances yet to see ourselves as belonging to one world, and not several" (Toye, 1987, p. 37). These considerations are essential to understanding how natural resources can become a commons in the near future, as well as the ethical underpinnings such project would require. For this reason, after having sketched out a map of the inequalities that characterise globalisation and the value placed predominantly by Western metaphysics on the separate notions of "nature" and "culture" it upholds, this chapter will attempt to conceptualise a commons through diversity, pointing to the ethical changes required to make this possible. Thinking of global politics through the lenses of cosmology and epistemology may, in fact, shed light on the ethical and political possibilities of an alternative approach to the commons, one which recognises the interdependence of all things.

Our worlds, or "*cosmoi*", are created through a contingent, open and ever-changeable conglomerate of axioms representing what we consider to be true (Campagna, 2019). These are constellations of cultural, social, historical, mythological, spiritual and indeed political principles that have upheld the world-view of different cultures, including how they conceive of the cosmos and how they act within it. For instance, the world of the Runa in Ecuador (indigenous group also known as Quechua) is pervaded by the divine incarnation of nature, or La Pachamama (represented below in a seventeenth-century illustration, Fig. 1), which will consequently create a way of seeing oneself and other beings as interconnected as opposed to the traditional Christian cosmology, for example. This Western cosmology, on the other hand, can be seen in an illustration from the same era of Diego Valadés, a Mexican-born cleric and author of the "Great Chain of Being" (Fig. 2), where a hierarchical understanding of the world departs from the heavens, reaching down firstly to humans, flying, swimming and land creatures, plants and, finally, non-living beings. In the words of Yuk Hui, cosmologies in this sense are "schemas that define modes of participation, but also correspond to the moral grounds of such participation" (2017). To summarise, ethical thinking is profoundly embedded in the world-view that one upholds.

In a *Playboy* interview, Kurt Vonnegut expressed these ideas in a more rudimental way, in the recognition that "there are thousands of other cultures and they all work pretty well; that all cultures function on faith rather than truth; that there are lots of alternatives to our own society" (Allen, 1988). Reality in this sense, however, is not

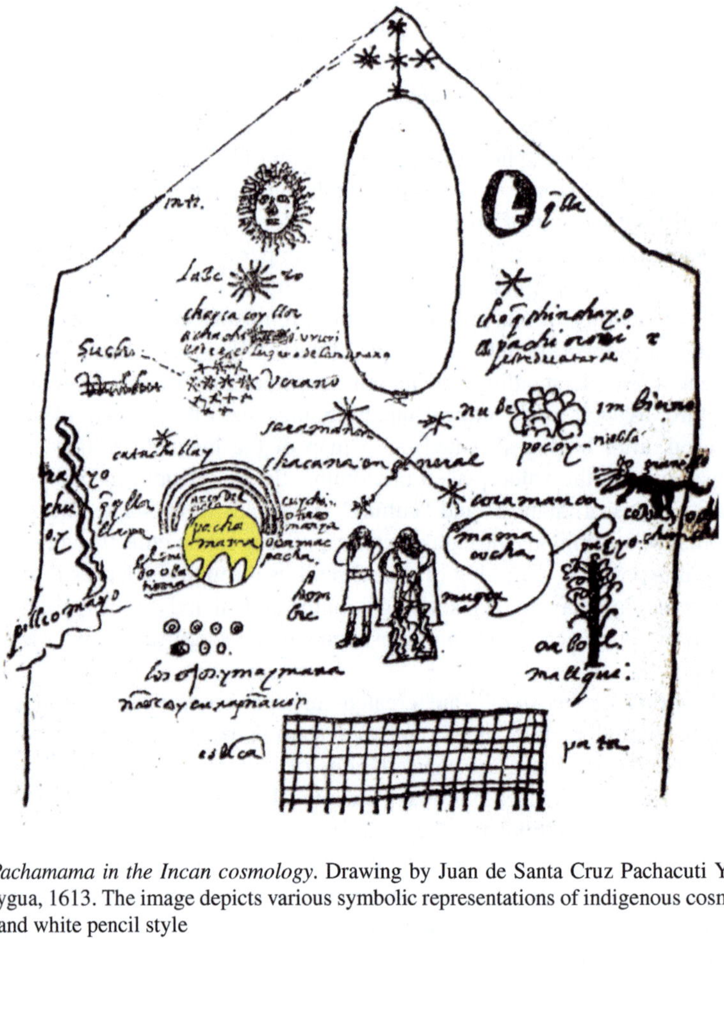

Fig. 1 *Pachamama in the Incan cosmology.* Drawing by Juan de Santa Cruz Pachacuti Yamqui Salcamaygua, 1613. The image depicts various symbolic representations of indigenous cosmology in black and white pencil style

Fig. 2 *Great Chain of Being*. Drawing by Diego Valadés, 1597. The image depicts an old black and white drawing of heavenly, god-like and angelic figures

merely observed from a multiplicity of perspectives in relation to a "monopolistic vision of truth" (Mol, 1999) but it is also created, performed, enacted. How different cultures associate themselves with natural resources, for instance, will be determined by the role and value these are given within their way of conceiving and creating their world. The definition of the Zapatistas summarises this well, as "a world in which many worlds fit" (EZLN, 1996). This view has also been embraced by the anthropologists of the ontological-turn movement, who aim to tackle the issue of modernity through the recognition of these ontological differences, such as Philippe Descola, Arturo Escobar and Eduardo Viveiros de Castro, as well as proponents of the multispecies movement, represented first and foremost by Donna Haraway. These two schools of thought share an urgent call for a renewed relationship between human and non-human entities, yet can be said to be divided by a "preference" respectively for culturalism and naturalism (Hui, 2020).

If one apprehends themselves as being part of nature rather than separate from it, this belief will reflect not only a different moral stance or representational view, but will radically shape their world and how one acts within this. This relationship will not mount to a mere "*inter*-action" between (human) culture and (more-than-human) nature, thus between pre-determined and separate bodies that act with each other, which presupposes a metaphysics of individualism, but rather will take the shape of an "*intra*-action" occurring within nature itself, where subject and object do not pre-exist but are ontologically inseparable and emerge with these (Barad, 2007). In this sense, this relational ontology advocates "a relationality between specific material (re)configurings of the world through which boundaries, properties, and meanings are differentially enacted and specific material phenomena" (Barad, 2007, p. 139) in a way that will allow one to account for how distinctions such as "nature" and "culture" are produced and the consequences they entail.

According to the anthropologist Eduardo Viveiros de Castro, this notion of culture, and therefore humanity, as part of nature rests at the heart of the concept of Amerindian "perspectivism". This term reflects the various conceptions founding the cosmologies of Amazonian indigenous peoples wherein the notions of "nature" and "culture" cannot be reduced to the dichotomy of Western thought, where the opposition of universal and particular, body and mind, necessity and spontaneity, animality and humanity are counterposed. In understanding that human cultures are "fundamentally in conflict over the mode of symbolisation they (conventionally) privilege as an element appropriated for action or intervention, in reserving to the other the function of the 'given'" (Viveiros de Castro, 2014, p. 53), it becomes clear that what one may define as nature turns out to be culture for another. The import of Amerindian perspectivism, in this sense, is to open up the potential for multiple visions beyond the human. In fact, it recognises "a spiritual unity and a corporeal diversity" or "a representational or phenomenological unity which is purely pronominal or deictic, indifferently applied to a radically objective diversity" where "culture or the subject would be the form of the universal, whilst nature or the object would be the form of the particular" for "a perspective is not a representation" (Viveiros de Castro, 1998). In this sense, a notion of "multinaturalism" is counterposed to the traditional western "multiculturalism", therefore recognising the presence of a common human and cultural condition which is incarnated through different corporeal states. When asked if culture is an intervention on nature, in fact, the indigenous Brazilian activist Aílton Krenak replied: "No, nature is an invention of culture" (Meirelles, 2020).

It can be argued (de la Cadena & Blaser, 2018; Escobar, 2018; Moore, 2015) that the issue underlying the poor, and extractive, treatment of natural resources (or, the commons) is a product of the Western separation between nature and culture which underpins the logic of industrialism, extractivism and capitalism, namely those activities which are putting at risk the availability and health of the environment for all. More generally, however, it can be attributed to the systematic dismissal of certain cosmologies in favour of others, particularly those belonging to and reproduced in the Western world which are often entangled in webs of colonialism, racism and sexism (Nirmal, 2020). Herein, nature is often characterised as "standing-reserve" (Heidegger, 1954, p. 14) (Fig. 3), thus as isolated from humanity, which in the 'Great Chain of Being' is above it, merely waiting to be harnessed and exploited as such. This is not the product of sheer extractive and exploitative practices, but rather the reflection of a collective mindset that governs the logic of Western thought. The question that arises is therefore this, how do we govern a commons, both material, such as an ocean or forest, and immaterial, such as knowledge or language, within which multiple worlds, and their respective understandings of "nature" and "culture", exist and interact with each other? More specifically, how do we envision a commons that is shared and contended amongst a constellation of plural cosmological, epistemological and ethical values?

Fig. 3 *Nature as "standing-reserve": an example of Western cosmology at play*. California, US, 2022. The image depicts a cluster of oil extraction rigs in a desert setting

As we currently find ourselves in the midst of several overlapping crises, of which natural resource and energy can be said to be the most prominent due to the dependence of ecosystems and human health upon these (Bansard & Schröder, 2021), understanding which worlds count helps us unveil the injustices, as well as create the pathways for future solutions that can help solve these. A recognition of the fact that there has been a historical dismissal of certain cosmologies, or worlds, deserves fundamental reflection when conceptualising the commons, particularly as the commons developed independently as a form of property amongst various early communities across the world (Wall, 2014). Along with this recognition, the task requires a form of humility, a term derived from the latin *humus*, the earth itself, to recognise the multiple worlds that are contended, stolen and owed within the one we share.

3 On Commons and Uncommonalities: Beyond Dichotomies

When we speak of the commons, questions inevitably arise around commonalities and uncommonalities. The "commons" as a "social system for the long-term stewardship of resources" also includes the social norms and practices (or "commoning") needed for their management (Bollier, 2011). It is important to note in fact that the commons are embedded within larger environments that also comprehend social relations and other systems (De Angelis, 2014). The commons are in fact necessarily produced and sustained by communities, which can act in local and trans-local "places" through social networks of mutual aid, solidarity and exchange practices that cannot be reduced to market form (De Angelis, 2003). In other words, there cannot be a commons without community. The often the inflationary use of this term, particularly referring to an abstract community, may risk "becoming a safety valve for capitalism" such as in the World Bank employment of the term "global commons" (1998) to indicate the mere privatisation of forests, seas and other resources (Federici, 2020; Rodotà, 2016). The benefit of this notion for larger, global scales in order to protect resources from these forces of market enclosure, liberalisation and privatisation will require new types of federation, laws, governance and culture. Within this, however, it requires new ways of understanding the particular ontology that is attempting to transform all worlds into a single universal one (Escobar, 2015) as well as an appreciation that the commons reflect "an interest in common" which however is not "the same interest" for the various cultural forms and knowledges involved (de la Cadena, 2015).

Epistemological diversity can be defined as "the diversity of knowledge systems underlying the practices of different social groups across the globe" (de Sousa Santos et al., 2008, p. 20). This is essential to the management of the commons as it underpins other sorts of diversity. Particularly in the context of the Global North, this has gained epistemological privilege through the establishment of modern science and positivism as a "higher" form of knowledge. The sociologist Boaventura de Sousa Santos has coined the term "epistemicide" to describe the suppression of other knowledges, a practice that was foundational to the model that the Global North has exported abroad (2017). This is a reflection of political, and ethical, struggle through which contestations in resources, territories, political power have grounded the struggle for what is to be considered as rational knowledge (Haraway, 1988). In other words, this epistemological "monoculture" that rests at the heart of the globalised world-view is one that has battled throughout history to establish itself as the dominant way of knowing and thinking. Nonetheless, as a reflection of the Global North's world-view, it is not one that should be relegated to the side, but rather coexist with other forms of knowledges to gain from the vast and localised diversity of human knowledge, as well as better inform the management of common resources in a context of environmental crisis.

All knowledge that is not strictly scientific, can be said to be "subaltern", or non-hegemonic, in Gramscian terminology. The notions of "hegemony" and "subalternity" are found in Antonio Gramsci's "Some aspects of the southern question" (1978), where he discusses the divide between the proletariat of the industrialised north and the rural south in Italy. These have subsequently been adopted by other intellectuals, particularly by postcolonial studies and the Indian branch of Subaltern Studies (see Spivak, 1988), to describe a wider global landscape. This intellectual movement has attempted to subvert the notion that traditional historiography has constructed of a lacking, underdeveloped "South" and a diametrically opposite "North" that must be imitated in its course of development. Similarly, this notion has been taken to understand not only the internal dynamics that may exist in the form of northern/southern, rich/poor, industrial/rural divides within nations of the global North itself, but also to describe global phenomena, such as colonialism and imperialism, as the products of historical relationships between power and knowledge. The production of a colonial discourse on behalf of nations from the Global North, devised to dehumanise and hegemonise the Global South, is an example of this (Said, 1978). The "southern question" can therefore be seen as a discursive construction (Petrusewicz, 1998) to subordinate the South and justify the North's material exploitation and its imposition of socio-economic standards following the North-European and Atlanticist model. To give voice to the subalternity of non-western, third-world histories and experiences, on the other hand, can help create new discourses by "provincializing Europe", the hegemonic imaginary figure which has vastly overshadowed habits of thought in other regions of the globe (Chakrabarty, 2008). These alternative narratives, in fact, are brought together by an attempt to uncover the ways in which the "South" has been represented as "other", or as "reverse" through acts of epistemic violence (Conelli, 2013).

These considerations are useful to reconceive of a commons through diversity. This will require the relocalisation not only of the commons and the institutions governing them, but also of social, productive and cultural activities spanning from food systems to learning, dwelling, healing and utilising energy, as well as an effort to recommunalise social life in a way that values the relational aspects of existence (Escobar, 2022). These transformations must be attuned to the particularities of local people and places, thus respecting their plurality in a renewed epistemological diplomacy. Hannah Arendt, in fact, writes that "plurality is the condition of human actions because we are all the same, that is, human, in such a way that nobody is ever the same as anyone else who ever lived, lives, or will live" (1958). Along with this plurality, is the recognition that the world is also a shared one amongst many who often hold different views of it. For this reason, echoing Arendt again, "living in a real world and speaking with one another about it are basically one and the same" (2005). Differences in ways of knowing, understanding, governing this shared world therefore become complementary, rather than fracturing the world in a manichean manner, thus leading to the creation of irreducible dichotomies between North and South, between Self and Other, between local and global. From a material and geographical point of view, this is the point Doreen Massey makes when she claims that space is the product of interrelations, particularly of coexisting heterogeneity, and is always under construction (2015). In other words, to preserve a common world, which includes both its reality and humanity, it is fundamental to preserve the plurality of opinions and open up the possibilities for conversation and contestation across uncommonalities.

Given this context, with its plurality of worlds and knowledges, it becomes clear that the pressing issue with the commons is not merely around how to manage these shared resources, but rather what meaning these may be given: this, as a collection of multiple understandings, interpretations, values and norms, is the framework within which decisions concerning the governing of commons are made. For this reason, this pursuit requires institutions that allow for the coexistence of subjects and practices that are diverse, multiple and in constant confrontation, to find forms of collaboration that are founded not on homogenisation but on multiplicity (De Angelis, 2022). In fact, "no formation of opinion is ever possible where all opinions have become the same" (Arendt, 1963). In this alternative sense, diversity is does not merely become reduced, but rather is preserved and intertwined to create new collective subjectivities, shared heterotopies, spaces where uncommonalities meet. The final section of this chapter will turn to more political implications of this discussion, to shed light on ways in which a commons can be reimagined in practice.

4 Where Worlds Meet: Commons in Practice

In the contexts of the many crises we are facing today, including the energy, environmental and food crises, practicing the commons becomes an increasingly urgent task. In order for this to happen, however, I argue that we must overcome the cosmological limitations of the current economic regime, the driving force of the Global North, to reimagine ways in which property, ownership and hierarchies are enacted and understood. In this way, the commons becomes a boundary object bringing together multiple values, practices, ways of thinking, seeing and knowing to extend the political imagination in a planetary way.

As the renowned declaration of Fredric Jameson goes, "it seems easier for us today to imagine the thoroughgoing deterioration of the earth and of nature than the breakdown of late capitalism; perhaps that is due to some weakness in our imaginations" (1994). In fact, the world is characterised by societies that have depleted their resources, distributed wealth unequally, degraded their landscapes will continue to do so as an alternative to solving the challenges they are faced with and a way to maintain present structures in place (Rolston III, 1996). More specifically, the highly unequal consumption rates prompted by neoliberal societies, such as the United States, has made it so that in order for every person across the globe to live like an American, five Earths would be required (EOD, 2018). As Viveiros de Castro and Danowski write, "that means that the country to the north of Mexico owes the rest of the world at least four worlds, in an unexpected transformation of the mythical humans without world theme. [...] The problem above all is that there are too few people with too much world, and too many people with way too little" (2018). This becomes evident in the simple fact that though poorer countries supply a great amount of raw materials to richer ones, their revenues remain significantly lower than the latter predominantly because of "the positions occupied in global supply chains and their respective roles in the world economy (Dorninger et al., 2021). To address the practices of overconsumption, overproduction, overexploitation of natural resources and to avoid the so-called "tragedy of the commons" implies a more profound question around the possibilities of moving beyond the logic of the market economy, given this context. The notion of a commons to be governed outwith the limitations and standards of neoliberal structures will, in fact, require a strengthening of the political imagination in order to propose new ways of thinking about natural resources, private property and exploitation, as well as new modes of organisation, socialisation and governance amongst "commoners".

From the 1980s, the rise of rampant neoliberalism was promoting enclosures of "commons" from community land to water resources, from education to urban spaces, energy development projects and protected conservation areas. According to the Midnight Notes Collective, these can be called the "new enclosures" (1990). Globalisation, more generally, can be described as a war against anything that is collective or communal, promoting in this way the ontology of the individual as severed from community (Escobar, 2022). The appropriation of common resources such as land, rivers, mountains, coasts, as well as human institutions, for their enclosure and capitalist conversion into an economic resource is happening day in and day out: the case of the Narmada Valley Dams project in India (Khan, 2012), the slow privatisation of National Health Services in the UK (Fisher, 2021), the controversial delimitation of marine protected areas in Antarctica (Smith et al., 2016), the legal enclosure of territories belonging to the Saami people for resource extraction in Norway (Anaya, 2011) are only some examples. These new enclosures are often characterised by the subordination of local or indigenous laws to private governance managed by corporations or state authorities, thus proving that ways of life that are more in touch with the surrounding landscapes (as are many indigenous populations or coastal communities, for instance) are dismissed on the grounds of extraction of resources or incremental economic profits.

In a similar way, however, there is an opposite movement that day in and day out is contributing to the "commoning" of resources and public life at a local scale: community gardens, occupied social centers, autonomous spaces and community energy projects are only some of the initiatives that are often moving beyond capitalist relations through the collectivisation of resources and an openness to plurality. A more institutionalised and explicit example of this is the Bologna Regulation for the Care and Regeneration of Urban Commons, adopted by the city in 2014 with the purpose of reconceptualising how government and citizens may cooperate to manage and design public spaces (LabGov, 2016). The commons have been found useful to foster the behaviours, responsibilities and democratic approaches that are needed to produce active citizenship within individuals (Rodotà, 2016). These practices are not only changing how the people involved relate and interact with one another, but also challenging the notion that people's needs can only be established through bureaucratic institutions, governments, urban planners and lawyers (Bollier & Helfrich, 2015). According to Stavrides, there are two ways of conceiving of the commons, one based on community and one on the public. While the first refers to a homogenous group of people who may share a similar identity and values, the latter puts emphasis on the relationship between different communities, as a space of diverging ways of life and values on a larger scale. Therefore, while a commons based on community would promote a ground of "affirmation of what is shared", the notion of the public can help establish a ground of negotiation that does not shy away from the difficulties, conflicts and contestations involved in sharing a commons (2010). This would not merely become symbolic of a different type of organisation or governance, but rather of a different practice of politics altogether, namely one that cuts across diversity and divergence, namely what Isabelle Stengers (2011) describes as practices of becoming while preserving heterogeneity, through an interest in common that need not be the same interest.

To conclude, as the context that has been described so far is one of profound multiplicity and diversity, an attitude of openness and deliberation is essential to ensure that the new institutions, processes and forms of organisation may not merely reproduce traditional, hierarchical forms of political organisation, but rather be constructed to reflect the values and ideals that are held by the commoners. Rather than serving a merely utilitarian function of stewardship, this commons would represent a coexistence of many worlds despite their cosmological, epistemic or cultural divergence. By being able to account for common interests, although they may not reflect the *same* interests, the cooperation of diverse commoners could call into question the traditional notions of power and knowledge, and alter the course of capitalist forces that are continuously ploughing their way towards environmental catastrophe. This alliance of ontological multiplicities, in the holding of a singular commonality, would ignite the imagination needed to discover new political pathways and innovative means of organisation.

To distil the above discussion, the main lessons towards a diverse ethics of the commons drawn from this chapter can be summarised as follows:

- Commons, both material and immaterial, are boundary objects able to bridge the plural worlds which constitute our reality and meet through what is shared.
- There has been a dismissal of certain world-views throughout human history: understanding which worlds count can help us create just pathways for future commons.
- An ethical approach to the commons requires a profound recognition of cosmological pluralism and embraces the plurality that this entails (be it epistemic, linguistic, technological etc.).
- Moving beyond exploitative relations and towards the collectivisation of resources and well-being can be achieved through an ethical management of the commons.
- Commons reflect ethical practice when they explore the commonalities rather than the divergences in our cosmologies.

References

Allen, W. R. (1988). Kurt Vonnegut: Playboy interview, July 1973. In *Conversations with Kurt Vonnegut*. University Press of Mississippi.

Anaya, J. (2011). *The situation of the people in the Sápmi Region of Norway, Sweden and Finland*. Human Rights Council, UN Doc. A/HRC/18/35/Add.2.

Arendt, H. (1958). *The human condition*. University of Chicago Press.

Arendt, H. (1963). *On revolution*. Viking Press.

Arendt, H. (2005). Introduction *into* politics. *The promise of politics* (pp. 93–200). Schocken Books.

Bansard, J., & Schröder, M. (2021). *Brief #16 the sustainable use of natural resources: The governance challenge*. Available at: https://www.iisd.org/system/files/2021-04/still-one-earth-natural-resources.pdf

Barad, K. (2007). *Meeting the universe halfway: Quantum physics and the entanglement of matter and meaning*. Duke University Press.

Bollier, D. (2011). *Commons, short and sweet*. Available at: https://www.bollier.org/commons-short-and-sweet

Bollier, D., & Helfrich, S. (2015). *Patterns of commoning*. The Commons Strategies Group in cooperation with Off the Common Books. Available at: https://patternsofcommoning.org

Campagna, F. (2019). *Technic and magic*. Bloomsbury Academic.

Chakrabarty, D. (2008). *Provincializing Europe: Postcolonial thought and historical difference*. Princeton University Press.

Chapin, F. S., III, et al. (2011). Earth stewardship: A strategy for social-ecological transformation to reverse planetary degradation. *Journal of Environmental Studies and Sciences* (1), 44–53.

Conelli, C. (2013). *Mezzogiorno postcoloniale*. Eleaml. Available at: https://www.eleaml.org/rtfnofor/2013_pdf/2013_05_15_conelli_mezzogiorno_postcoloniale_nola_13_aprile.pdf

De Angelis, M. (2003). Reflections on alternatives, commons and communities or building a new world from the bottom up. *The Commoner* (6). Available at: https://files.libcom.org/files/1_deangelis06.pdf/

De Angelis, M. (2014). Social revolutions and the commons. *South Atlantic Quarterly, 113*(2), 299–311.

De Angelis, M. (2022). *Spazio comune. città come commoning—recensione di Massimo De Angelis, Effimera*. Available at: https://effimera.org/spazio-comune-citta-come-commoning-di-massimo-de-angelis/

De Angelis, M., & Harvie, D. (2014). The commons. In M. Parker, G. Cheney, V. Fournier, & C. Land (Eds.), *The Routledge companion to alternative organizations* (pp. 280–294). Routledge.

de la Cadena, M. (2015). *Earth beings: Provincializing nature and the human through Andean worlds*. Duke University Press.

de la Cadena, M., & Blaser, M. (2018). *A world of many worlds*. Duke University Press.

de Sousa Santos, B. (2017). *Epistemologies of the South*. Routledge/Taylor & Francis.

de Sousa Santos, B., Nunes, J. A., & Meneses, M. P. (2008). *Another knowledge is possible: Beyond northern epistemologies*. Verso.

Dorninger, C., Hornborg, A., Abson, D. J., von Wehrden, H., Schaffartzik A., Giljum, S., Engler, J.-O., Feller, R. L., Hubacek, K., Wieland, H. (2021). Global patterns of ecologically unequal exchange: Implications for sustainability in the 21st century. *Ecological Economics, 179*. https://doi.org/10.1016/j.ecolecon.2020.106824

Earth Overshoot Day. (2018). *How many earths? How many countries?*. Earth Overshoot Day. Available at: https://overshoot.footprintnetwork.org/how-many-earths-or-countries-do-we-need/#:~:text=The%20Ecological%20Footprint%20for%20the,if%20everyone%20lived%20like%20Americans

Escobar, A. (2015). Degrowth, postdevelopment, and transitions: A preliminary conversation. *Sustainability Science, 10*(3), 451–462. https://doi.org/10.1007/s11625-015-0297-5

Escobar, A. (2018). *Designs for the pluriverse: Radical interdependence, autonomy, and the making of worlds*. New Ecologies for the Twenty-First Century. Duke University Press.

Escobar, A. (2022). On the ontological metrofitting of cities, where is here? *eflux Journal*. Available at: https://www.e-flux.com/architecture/where-is-here/453886/on-the-ontological-metrofitting-of-cities/

Federici, S. (2020). *Revolution at point zero: Housework, reproduction, and feminist struggle*. PM PR.

Fisher, A. (2021) The NHS is being privatised by stealth under the cover of a pandemic. *The Guardian*. Available at: https://www.theguardian.com/commentisfree/2021/may/03/government-pandemic-privatise-nhs-by-stealth

Gramsci, A. (1978). 'Some aspects of the Southern question', *Selections from political writing: 1926* (Q. Hoare, Trans.). International Publishers.
Haraway, D. (1988). Situated knowledges: The science question in feminism and the privilege of partial perspective. *Feminist Studies, 14*(3), 575.
Heidegger, M. (1954). *The question concerning technology*. Garland Publishing, Inc.
Hui, Y. (2017). On cosmotechnics. *Techné: Research in Philosophy and Technology, 21*(2), 319–341.
Hui, Y. (2020, December). *For a planetary thinking—Journal #114*. e-flux. Available at: https://www.e-flux.com/journal/114/366703/for-a-planetary-thinking/
Jameson, F. (1994). *The seeds of time*. Columbia University Press.
Järvensivu, P., et al. (2018). *Global sustainable development report 2019: Governance of economic transition*. Available at: https://bios.fi/bios-governance_of_economic_transition.pdf
Khan, T. R. (2012). "Dam" the irony for the greater common good: A critical cultural analysis of the Narmada dam debate. *International Journal of Communication, 6*, 194–213.
LabGov. (2016). *Bologna regulation on public collaboration for urban commons*. LabGov. Available at: https://labgov.city/theurbanmedialab/bologna-regulation-on-public-collaboration/
Massey, D. (2015). *For space*. SAGE.
Meirelles, M. (2020). *Our worlds are at war, Journal #110*. Available at: https://www.e-flux.com/journal/110/335038/our-worlds-are-at-war/
Midnight Notes Collective. (1990). *The new enclosures*. Available at: http://www.midnightnotes.org/newenclos.html
Mol, A. (1999). Ontological politics. A word and some questions. *The Sociological Review, 47*(1 Suppl), 74–89.
Moore, J. W. (2015). *Capitalism in the web of life: Ecology and the accumulation of capital*. Verso.
Nirmal, P. (2020). Nature and gender. In A. Kobayashi (Ed.), *International encyclopedia of human geography* (2nd ed., pp. 285–293). https://doi.org/10.1016/B978-0-08-102295-5.10232-X'
Petrusewicz, M. (1998). *Come il Meridione divenne una Questione. Rappresentazioni del Sud prima e dopo il Quarantotto*. Rubettino.
Rodotà, S. (2016). *Beni comuni—bozza*. Available at: https://www.mbsconsulting.com/wp-content/uploads/2016/07/Beni-comuni-di-Stefano-Rodota.pdf
Rolston III, H. (1996). Feeding people versus saving nature? In W. Aiken & H. LaFollette (Eds.), *World hunger and morality* (2nd ed., pp. 248–267). Prentice-Hall.
Said, E. (1978). *Orientalism*. Pantheon Books.
Smith, D., McGee, J., & Jabour, J. (2016). Marine protected areas: A spark for contestation over 'rational use' of Antarctic marine living resources in the Southern Ocean? *Australian Journal of Maritime & Ocean Affairs, 8*(3), 180–198. https://doi.org/10.1080/18366503.2016.1229398
Spivak, G. C. (1988). Can the subaltern speak? In C. Nelson & L. Grossberg (Eds.), *Marxism and the interpretation of culture* (pp. 271–313). Macmillan Education.
Stavrides, S., De Angelis, M. (2010). *On the commons: A public interview with Massimo De Angelis and Stavros Stavrides, Journal #17*. Available at: https://www.e-flux.com/journal/17/67351/on-the-commons-a-public-interview-with-massimo-de-angelis-and-stavros-stavrides/
Stengers, I. (2011). Comparison as a matter of concern. *Common Knowledge, 17*(1), 48–63.
Tomlinson, B. R. (2003). What was the third world? *Journal of Contemporary History, 38*(2), 307–321. https://doi.org/10.1177/0022009403038002135
Toye, J. (1987). Dilemmas of development: Reflections on the counter-revolution in development theory and policy. *Southern Economic Journal, 56*(1), 258. https://doi.org/10.2307/1059077
Viveiros de Castro, E. (1998). Cosmological deixis and Amerindian perspectivism. *The Journal of the Royal Anthropological Institute, 4*(3), 469. https://doi.org/10.2307/3034157
Viveiros de Castro, E. (2014). *Cannibal metaphysics*. University of Minnesota Press.
Viveiros de Castro, E., & Danowski, D. (2018). Humans and terrans in the Gaia war. In M. de la Cadena & M. Blaser (Eds.), *A world of many worlds* (pp. 172–204). Duke University Press.
Wall, D. (2014). *The commons in history: Culture, conflict, and ecology*. OUP Academic. Available at: https://academic.oup.com/mit-press-scholarship-online/book/20160

World Bank. (1998). *Protecting the global commons | world development report 1999/2000.* Available at: https://elibrary.worldbank.org/doi/abs/10.1596/0195211243_Chapter4
Zapatista Army of National Liberation. (1996). *EZLN: Fourth declaration of the Lacandon Jungle.* Available at: http://struggle.ws/mexico/ezln/jung4.html

Erin Rizzato Devlin is affiliated with the Center for Energy Ethics at the University of St Andrews in Scotland, and co-founder of Equitable Energy Research CIC, a non-profit dedicated to justice in the green transition. Her research interests lie in energy politics, sustainable governance, human rights and environmental justice, particularly in rural and island contexts. She collaborates on interdisciplinary projects addressing energy transitions and the ethical tensions therein.

Open Access This chapter is licensed under the terms of the Creative Commons Attribution-NonCommercial-NoDerivatives 4.0 International License (http://creativecommons.org/licenses/by-nc-nd/4.0/), which permits any noncommercial use, sharing, distribution and reproduction in any medium or format, as long as you give appropriate credit to the original author(s) and the source, provide a link to the Creative Commons license and indicate if you modified the licensed material. You do not have permission under this license to share adapted material derived from this chapter or parts of it.

The images or other third party material in this chapter are included in the chapter's Creative Commons license, unless indicated otherwise in a credit line to the material. If material is not included in the chapter's Creative Commons license and your intended use is not permitted by statutory regulation or exceeds the permitted use, you will need to obtain permission directly from the copyright holder.

Ethical Commoning: A Framework for Identifying and Comparing Values, Practices, and Capacities

Stefan Partelow, Guido Caniglia, and Christopher Luederitz

Abstract The term commoning has acquired a variety of meanings. However, despite the intuitions that many scholars have that commoning has a normative orientation aligned with community building, justice and sustainability, there are no clear conceptual frameworks detailing the values and ethics of commoning practices. Such frameworks would enable research on commoning to descriptively and normatively identify and document its practices, while differentiating commoning from other less or non-desirable collective action and community building and collective action activities (e.g., organized crime; anti-democratic or discriminatory political movements; collusion). Current research has focused on community-level collective action processes that embody, enable, create or sustain resources held in common, that is the commons. Others emphasize the practices and performances of becoming in common as commoners, entire communities, and their environment. Despite these differences, similar values often underpin commoning—such as solidarity, fairness, care for the environment—but which have yet to be synthesized into a working framework. In this chapter, we ask: what does an ethics of commoning look like? How can we account for what it means to do commoning (ethically) well? But also, how can we provide orientation to commoning? To address these questions, we introduce three interwoven components of a prospective ethics of commoning framework: (a) values of commoning, directed towards social justice and environmental sustainability; (b) alternative valuation practices, that engage with the materiality of the world outside of dominant ones; and (c) individual and collective capacities for commoning, based

S. Partelow (✉)
Center for Life Ethics, University of Bonn, Bonn, Germany
e-mail: stefan.partelow@uni-bonn.de

Faculty of Agricultural, Nutritional and Engineering Sciences (AEI), University of Bonn, Bonn, Germany

G. Caniglia
University of Helsinki, Helisinki, Finland
e-mail: guido.caniglia@helsinki.fi

C. Luederitz
Département de management, Université du Québec à Trois-Rivières, Trois-Rivières, Canada
e-mail: Christopher.Luederitz@uqtr.ca

© The Author(s) 2026

S. Partelow (ed.), *Ethics and the Commons*, Schriften zu Leben und Gesellschaft - Series on Life and Society 3, https://doi.org/10.1007/978-3-032-24246-4_4

on practical wisdom and virtue ethics. We use the framework to articulate and analyze the ethical dimensions of commoning. Through the idea of commoning wisdom, we use the framework to provide orientation and inspiration about how to think about the ethical work embedded in the practice of commoning.

1 Introduction

Communities are essential for human life, and act as the cornerstones of social, economic and political organization. Yet, most narratives of societal transformation have overlooked the community-level in favor of framing sustainability problems as needing to be solved by individual behavior change, private organizations (e.g., business practices) or state-level government regulation and intervention. While those levers are critically important, communities and informal social organization processes have received far less attention as levers for change, even among non-place-based communities such as communities within private, state and inter-state or multilateral organizations, or online.

Commons scholarship has long examined the role of communities in stewarding physical or environmental shared resource commons (van Laerhoven et al., 2020). The goal of classic commons research was to sustain environmental commons and physical infrastructure. More recent focus, however, has shifted to include sustaining the social actions and practices of community formation, as a result of holding or enabling shared resources (e.g., commons or values) as a relational perspective. These social actions and practices are often referred to as commoning. We review different perspectives and definitions of commoning below, but it is important to note that commoning practices and performances can foster new relations and subjectivities, often contingent with the exercise of power (Nightingale, 2019; Partelow & Manlosa, 2023). It is thus important to focus on the becoming of commoners, on the practices that allow for the emergence of communities, including the conflicts and exclusions.

What is conceptually important, but not often discussed, is the need to distinguish commoning activities from other forms of collective action and community building that may not align with the assumed values and ethics we should care about if we are to achieve just, democratic, inclusive and sustainable societies. As such, commoning practices in the literature tend to be oriented around a shared set of values that motivate, support, and drive a normative orientation towards the creation of more just and sustainable futures. Examples of these values are solidarity, care for others, environmental stewardship, and responsibility. Such values permeate the everyday lives of individuals and communities engaged in governing the commons (governance) and becoming commoners (subjectivities). The ethics and values of these actions are quite different from, for example, the successful collective action and community building processes of organized crime, malevolent disinformation networks in media, or political movements advocating non-democratic rule or oppression. To date, much has been written about the descriptive organizational and political—or on the ontological and anthropological features—of the commons (Anderies & Janssen, 2016; Ostrom,

1990). Yet, less attention has been paid to the ethical dimensions of commoning, especially of commoning efforts motivated by social justice, democratic political processes and environmental sustainability, while also avoiding community formation and collective action that seeks to erode those efforts. We still lack overarching frameworks to understand how such values are mobilized through practices that build individual and collective capacities towards justice and sustainability. Thus, we ask: What ethical principles or theories are aligned with commoning? How does commoning shape certain values, especially around social justice and environmental sustainability? What does it mean to do commoning (ethically) well? But also, how can we frame and understand the ethical dimensions of commoning as (partially) distinct from their political dimensions?

In order to address these questions, we propose a framework that allows for descriptive accounting, and then for normative orientation on the ethics of commoning practices for sustainability. The framework is based on three interwoven dimensions: (a) values associated with commoning, directed towards social justice and environmental sustainability; (b) value practices, that engage with the materiality of the world outside of dominant ones, (c) individual and collective capacities for commoning, based on practical wisdom and virtue ethics. Before introducing the framework, in Sect. 2, we start by clarifying some important features of commoning as highlighted in recent literature. In introducing the framework in Sect. 3, we first clarify the ethical dimensions of the commons and then present its main interwoven dimensions of values, practices, and capacities. We then suggest that the idea of commoning wisdom may be a useful means of orientation and inspiration (normatively) to what it would mean to do commoning ethically well. We conclude in Sect. 4 by discussing potential avenues for further research on the ethics of commoning.

2 Communities, Subjectivities, and Commoning: Definitions and Potential

Commoning is not a new feature of society, but our interest in commoning is being revived given the concept's potential for better understanding the actions and practices of community-building towards social justice and environmental sustainability, that are place-based and adapted to local contextual needs and goals. Throughout human history, practices of sharing and collective action have been essential for physical and cultural survival (De Angelis, 2021). In current contexts, cooperation and coordination challenges across local-to-global levels and scales is an ever-present political, economic and cultural reality. More broadly, many efforts are rethinking the broader production of society and social relationships, counteracting dominant narratives and policies of globalization and individualism (e.g., the protagonist self in an undefined global community). Being part of clearly defined communities is essential for the social-psychological wellbeing of individuals, because it shapes individual identity, values and behaviors. Communities are arguably the drivers of informal institutional formation such as trust, norms, values and meaning, which are critical for functional societies, but which are impossible to construct as individuals or to regulate with formal governance.

One reason why communities lack attention in sustainability research, is because they are hard to define. They tend to have strong informal dimensions and are typically self-organized, self-governed, and heavily premised on voluntary social relationships. They often either overlap with, or are mismatched or interwoven with other formal social structures in politics, the economy and other dimensions of civic life. For example, research in the field community psychology has shown that individual outcomes such as belief formation, health, and education are strongly shaped by their communities (e.g., neighborhoods, clubs, social groups) and their surrounding social-ecological conditions. These communities can offer empowering conditions for individuals to explore and set beliefs, behavioral patterns, and/or information sharing and learning practices, which are linked to the community's group-based belief system, core activities, relational environment, opportunity role structure, leadership, and setting maintenance & change (Maton, 2008). Communities tend to form relationship structures around shared features among members, which may include shared interest, shared history, kinship, shared location, shared culture, shared problems, shared social status or role, or shared values, among others, often with multiple features combined. Most communities are self-governed, meaning that the members are self-organized and coordinated in some formal or informal way. In contrast to the way most environmental common-pool resources are understood, social communities are strengthened through use and engagement, rather than degraded. As such, they erode by neglect, or through the disentanglement of the unifying features. However, the emergence of communities is also characterized by conflicts, disagreements, power imbalances, and dynamics of inclusion–exclusion (Nightingale, 2019; Singh, 2017). Nonetheless, such tensions are often the result of establishing the key features of many strong communities, such as clear membership boundaries, kinship and reciprocity norms, and shared actions and/or values of collective interest.

2.1 Complementary Perspectives on Commoning

Evolved from the commons literature, commoning is gaining traction for understanding how self-organized communities produce and reproduce themselves. In doing so, beliefs and values are formed, and capacities are enabled—both implicitly and explicitly—towards shared goals and actions. The term commoning is attributed to being popularized by historian Peter Linebaugh, but has evolved into various interpretations. Two major research perspectives on commoning have focused, respectively, on the governance and on the more subjectivity-oriented aspects of commoning.

Table 1 Different perspectives in the literature on commoning

Governance-oriented perspectives	Subjectivity-oriented perspectives
• Leveraging collective capacities for action • Self-organization • Managing what is held in common (e.g., property) • Communicating and coordinating • Solving (social-ecological) problems	• Relating and connecting • Collective ways of being • Shared mindsets and values • Enabling the social conditions • Emergence and guidance by ethical commitments

2.1.1 Governance-Oriented Perspectives

One way is its association with sustaining resources held in common (e.g., open access, common-pool resources, public goods) or other goods under common property rights (Table 1). Definitions in the literature from this perspective include "commoning creates commons" or that commoning is "voluntary and inclusively self-organized activities and mediation of peers who aim at satisfying needs" (Euler, 2018). Nayak and Berkes (2011) argue that commoning is the process of converting a resource into a jointly shared one.

2.1.2 Subjectivity-Oriented Perspectives

Other interpretations position commoning as the production of subjectivities (Table 1), that is individual and collective "ways of perceiving, understanding, and relating to the world" (Singh, 2017). These include shared communities themselves. Federici (2011) states that commoning must be the production of ourselves as a common subject, as there is no commons without community. Nightingale (2019) argues that societal transformation should aim to do commoning, to become in common, rather than seeking to fix property rights (e.g., to establish and maintain common property rights), there is a need to prioritize relations of sharing and collective practices as the foundation, while also emphasizing that commoning implies the "ambivalent outcomes of the exercise of power", with its dynamics of inclusion and exclusion. De Angelis (2017) interprets commoning as "a form of social labor that has a direct relation to the needs, desires and aspirations of the commoners…[creating] conditions of resilience and self-organization". Further, by emphasizing the importance and role of affective and communicative relationships, Singh (2017) invites focusing on the individual and collective processes of becoming commoners, where "the commons and the commoners are co-constituted through intersubjective communication and affective relations" and entangled ecologies of nature (Singh, 2017).

Partelow and Manlosa (2023) recognize how the two perspectives can be—or perhaps are—interrelated. They note that commoning is the "actions by groups with shared interests towards creating shared social and relational processes as the basis of governance strategy." Here, the similarities are emphasized across the two orientations in the study of the commons. This is that social groups exist at an interface of ethical value formation and practical actions of commoning (e.g., governance). These processes exist on a plain with multiple magnetic centers related to how people come together in a community, and through affective relations with nature, as well as how and why communities work together, create and share values. Importantly, these processes will differ across communities with different goals and capacities, but are likely unified by their overarching ethical commitments linked to shared ideologies such as social equity (for society beyond in-group) and environmental sustainability.

2.2 *Understanding Commoning Beyond Mechanisms*

It is evident that commoning cannot only be explained by the mechanistic processes of how social groups self-organize and undertake collective action, but through better understanding the interplay between ethical values, goals and capacities—between the individual and collective practices. For example, there is a need to couple what has been learned and built-in collective action theory (e.g., Ostrom's Design Principles [Cox et al., 2010]: group size, group boundaries, rule alignment, sanctioning), with ethics and the evaluative criteria in the sustainability literature. We want to know how groups can effectively work together, but also with what means, and towards what ends. For example, towards what desirable range of sustainability and transformation values and goals?

This is important because we form communities in our societies for many types of reasons, but orientation towards social equity or environmental sustainability may be completely missing. Some communities may be highly effective at self-organizing, working together and creating shared beliefs and norms, but be completely misaligned or antithetical to, for example, equity or sustainability. Examples might include organized crime, self-interested political lobbying, economic or political collusion, military alliances or extremist political or religious movements. All of these groups and communities will have clear goals, shared beliefs, good leadership, group boundaries and sanctioning. Such groups can clearly work together well and coordinate actions based on Maton's (2008) community empowerment conditions, but are missing the guiding moral principles that provide either normative orientation or observations of practice communities that illuminate clear values.

Most communities in society, however, fall on the multi-dimensional spectrum between the extremes of organized crime and self-sustaining off-the-grid ecovillages. Communities are highly diverse across society with different economic, social, cultural and political goals—which are working together—that can be analyzed and learned from with a commoning lens. Furthermore, most people are part of multiple communities that co-shape individual and collective values, goals and practices. What we believe is useful, is to provide a framework that can help orient, compare and empirically analyze different commoning communities. Thus, we outline the key concepts and consideration towards developing such a framework.

3 A Framework for Ethical Commoning: Key Concepts

Commoning needs a guiding framework that specifies its relation to ethical theories, concepts and/ claims. This should include its core distinguishing features that differentiate it from other mechanistic social organization theories. Frameworks serve numerous functions, and are useful as both knowledge synthesis and positioning tools to orient scholarly engagement, and to connect tangential empirical and theoretical developments around a problem-area, concept or theory (Partelow, 2023). We argue, it is useful to have an initial or tentative framework to deepen, guide and provoke discussions on the ethics of commoning. Frameworks are metaphorical fires for scholars to huddle around, giving light and warmth to conceptual dark areas that motivate discussion, arguments and debate over our observations, models and ideas.

One way to build a framework is to synthesize insights from the commoning literature about the practices and performances that enable commoning (Partelow, 2023). Another place to look is towards existing ethical theories that allow for identifying closely aligned value-oriented practices (Caniglia et al., 2023; Partelow et al., 2024). By bringing together such insights, a framework for the ethics of commoning may account for how community-level social organization processes may lead to incremental changes towards social equity and environmental sustainability, or possibly larger societal transformations that change the structure and function of social, economic and political systems.

We believe three malleable and tightly interwoven components of ethical life are helpful: (a) values, (b) valuation practices, (c) capacities (individual and collective). Each of the dimensions depends on and supports the others. For example, values can only be realized through practices, which in turn allow for the development of capacities that can reinforce formation of values and ability to practice them. How the pillars are defined and specified will likely differ from situation to situation, and it should, and be expected to change and evolve. Furthermore, these three components always exist within the context of group structures and the broader social-ecological context. Acknowledging this is also necessary for an ethical commoning framework, but can take different forms.

In the following subsections we detail the three components of our suggested guiding framework for an ethics of commoning (Fig. 1). Each subsection outlines the context proposes the core pillar, but also attempts to define and specify certain features of it as suggestions. These specific suggestions are drawn from recent literature on commoning, sustainability and ethics. We believe the interpretation and specifics of each pillar can and should evolve. We do, however, argue that the current literature we draw on offers useful insights as a starting point.

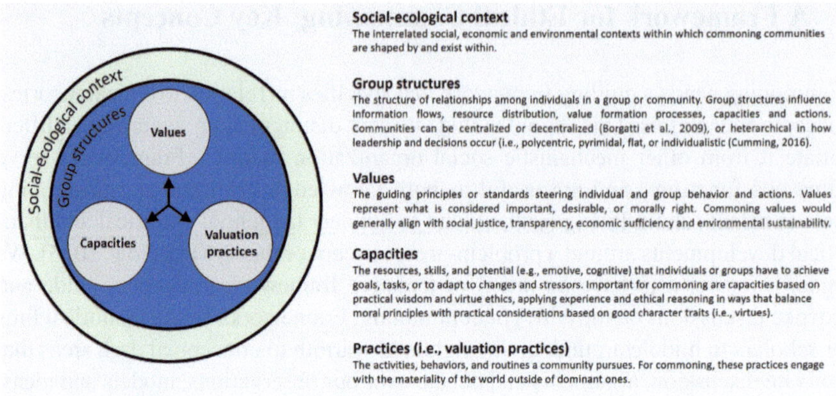

Fig. 1 A framework about the main components of ethical commoning. The three interacting features of ethical commoning intersect within the context of group structures, which exist within a broader social-ecological context. The 'Group structures' insights draw on work from Borgatti et al. (2009) and Cumming (2016). The image depicts a visual framework of concentric circles showing the embedded relationships between social-ecological context and group structures, with text to the right providing definitions of each terminology used

3.1 Commoning Ethics? Balancing Descriptive Observations and Normative Framings

Ethics can mean many different things. We believe building up an empirically observed and descriptive foundation is necessary. This should complement and balance our understanding and influence of normative ethics related to commoning. In doing so, we can begin to observe ethics as what Lambek et al. (2015) refer to as "a part of life as it is lived with others, worked out in interaction and conversation with others rather than transcendent of them, and within a given historical, cultural, and social context." The ethical dimension of living is not a fixed, identifiable body of facts, rules, reasons, or reflections. Rather, ethics are an emergent quality or property of life practices. Ethics is thus "embedded in or intrinsic to practical activity—as action in the world and in the ways in which we lead our lives" (Lambek et al., 2015). Ethics, as living and practical ethics, are placed within the lived praxis that is performed in and of the everyday (Kinnunen & Valtonen, 2017). Ethics is immanent to social life, that is to how people relate to one another in different contexts and situations, from daily and mundane interactions (e.g., how we talk to one another) to deliberations that are seen as having a strong ethical relevance (e.g., in relation to human and animal rights). Thus, addressing ethical questions means, first and foremost, in acknowledging and trying to account for the complexity, uncertainty and ambiguity of living together.

How can we attempt to capture the ethical dimensions of commoning as they are actually practiced and observed, while putting these descriptive observations in contrast and interrelation with the normative ethics of social justice and environmental sustainability? What are the immanent and experienced ethical dimensions of commoning? Addressing such questions will give us insight into the lived realities of commoning. This requires identifying and analyzing processes that cannot be reduced to their political (from less to more subversive, from more to less critical to existing powers) or institutional dimensions (from more organizational to collective action theory), which have been the focus of a lot of commons scholarship (Perperidis, 2024). Accessing the ethical dimension of commoning means to be able to ask questions about individuals' and communities' lived experiences of values, about their becoming in common as individuals and as collectivities (Nightingale, 2019; Singh, 2017). It means to look into the everyday lived encounters where ethical dimensions of commoning may manifest and come to the surface, included in their embodied, corporeal, affective, and emotional features (Diprose, 2002; Pullen & Rhodes, 2014).

This means to ask questions grouped around three broad categories such as *being, relating, and (re-)generating*: In relation to *being:* Who is the commoner (Singh, 2017)? What are their values and principles as reflected by actions? How do they cultivate such values and develop capacities? What kind of social and relational agents are they? How do they cultivate their own character and self, while in communication and relation with others? How do they react to or take action towards dynamics of marginalization and exclusion (Nightingale, 2019)? For *relating:* How do people decide to relate to one another (Bollier & Helfrich, 2015)? How do commoners become in common? What kind of practices do they engage with? How do they establish new relations with non-human and socio-nature relationships (as means towards ends or as ends in themselves)? What is their sense of dignity, for example? How does a community react or take action in relation to misconduct and even how is misconduct defined or framed? How do they forge collective and mutual rules and boundaries regarding ethical principles (Perpedis, 2024)? For *(re-)generating:* What do they feel responsible for, and how do they make decisions and take action in relation to this sense of responsibility? But also, how do they value their environment, their relationship to and understanding of nature, other people, and communities? What is their understanding of the natural resources they attempt to conserve or regenerate? What do individual and communities decide to attend to, prioritize and give attention to? What do they give value to? What values do they embrace and how do they enact those values? What kind of people are they or set out to be or aspire to be?

Asking such questions requires better understanding of the ways in which commoners and communities of commoning live. Neutrally observing commoning practices—rather than imposing an understanding of what is right or wrong, or what a good life is—can reveal diverse meanings and understandings of what we think we know commoning to be. To guide this analytical process of discovery, a framework is an important tool to balance inductive and deductive modes of learning and knowing through scholarly comparison, while also being open enough to help uncover novel and innovative ways of thinking about ethics and commoning.

3.2 Group Structure and Social-Ecological Context

Individuals and communities co-shape their ethics through dynamic interactions that form group structures. Group structure needs to be acknowledged as an evident feature of communities that shapes the ethics of commoning in the form of values, practices and capacities. Group or network structures further shape governance approaches through leadership, participation and inclusion dynamics that will likely vary based on group or network size (Fig. 2). It is also important to acknowledge that different groups or communities also co-shape the values of each other. In this chapter, we do not intend to give a full analysis of the many dimensions of community structure, but to acknowledge and pinpoint a few areas of importance for future research examining descriptive and normative features of commoning constellations.

The framework and its core conceptual features we propose, need to be considered within the context of how individuals relate to the different community groups they are a part of, and how those community groups interact with and co-shape each other. Within these relations, the tension between observed descriptive structures, and normative or political oriented practices needs careful attention. Formal structures are often laden with informal power dynamics, and oppositional groups often co-shape their normative positions, goals, values and leadership. This is particularly the case if different groups share common resources, where decisions, values and practices can have interdependent implications. For example, in climate politics, decisions taken by governance actors (i.e., civil society organizations, corporations, states) create the need for others to re-shape their positions, actions and values. There is a dynamic and iterative co-shaping of positions, values and/or organizational strategy based on what the other actors do or don't do. Yet, despite the many groups involved,

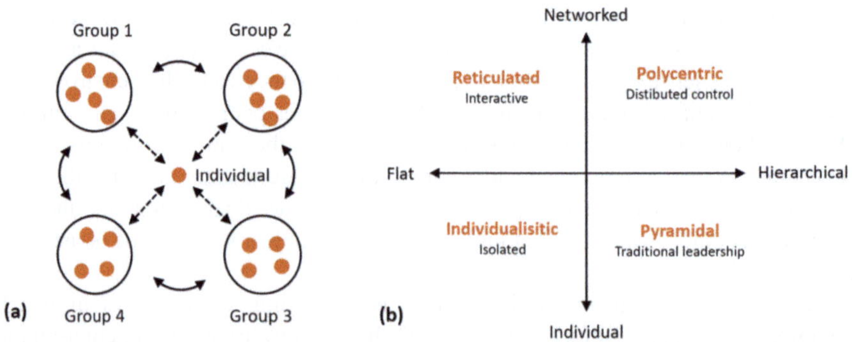

Fig. 2 (**a**) Individuals (i.e., orange dots) often participate in multiple social groups. Each group is likely to have different values, capacities and practices. Single or multiple individuals likely move fluidly between groups, co-shaping the values, capacities and values of each other. (**b**) Social groups have different types of leadership and decision-making structures, which will shape values, capacities and practices in commoning. Derived and modified from Cumming (2016)

climate politics and negotiations are centered around both a shared resource and commoning practices that evolve around recognition of shared interdependencies within and between our communities.

3.3 *The Role of Values*

Values are sets of beliefs that can motivate or guide actions. Values are central to ethics because they guide individuals' judgments about what is right and wrong to do. They provide orientation for designing and evaluating actions and their consequences. Values (e.g., those of transparency and fairness) help to establish consistent standards that can be applied across different situations. Furthermore, the environmental ethics and sustainability science literature has focused on the values of nature, such as intrinsic values (inherent worth of nature as an end in itself, regardless of any human interest) or relational values (that are human-centered; instrumental) (Eyster et al., 2023; Pratson et al., 2023). Instrumental values are relational values referring to those concerns related to the meaningfulness of relationships, such as those between nature and people and among people within nature or fostered by nature. A plurality of values is needed to understand the multiple ways in which nature matters to people (Horcea-Milcu et al., 2019). This plurality is essential when fostering new relationships to nature and the environment as "intrinsic, instrumental, and relational values are distinct but at the same time they are connected and can be simultaneously present when people articulate how and why nature matters to them" (Berta Martin-Lopez, 2020).

The values of commoning have been advanced more popularly by non-academic activists. This includes work from pracademics such as David Bollier, Silke Helfrich and Jay Waljasper, and countless community-led movements, particularly in the global south. This is most likely due to the association of commoning—even the emergence of the term—with practices and actions for sustaining commons, more than it has been an academic exercise. Table 2 presents examples of such values both as related to social justice and to environmental sustainability. These values shape commoning processes and constitute the baseline for understanding the main features ethics of commoning. Some values are universal (as related to basic human rights) and some are context specific (as emerging in specific cultures) (Centemeri, 2018; De Angelis, 2021). Although values are a core component of an emerging commoning framework and theory, what the specific values are is malleable, and is perhaps most effectively guided by a mix of descriptive observations of practices and theoretical positions. In developing a framework that accounts for the ethics of commoning we keep the opportunity open to observe new and diverse values that are translated and support the practices of commoning while also being principally anchored to certain pillars as a reference.

Table 2 Values, practices and capacities of ethical commoning observed by the authors of this chapter in the literature

Values	**Inclusion, participation and transparency**: Ensuring that all community members have voice, access and participation opportunity in decision-making and management **Responsibility and solidarity**: Recognizing shared care responsibilities while fostering mutual support and collective action **Proportionality and sufficiency**: Ensuring individuals use only what they need while contributing fairly based on their capacities **Social justice and equity**: Striving for fair access to resources and opportunities, while correcting imbalances and inequalities **Empowerment and autonomy**: Enabling self-organized decision-making that takes control over management without negative external coercion **Relational connections with people and nature**: Cultivating deep, reciprocal and respectful relationships in social-ecological systems **Aligning stewardship and use**: Leveraging resource use as a means to build care-based and mindful relations with nature that allow for sufficient social-ecological functioning over time **Social-ecological sustainability**: Managing the commons in ways that promote the resilience and health of social systems and ecosystems, ensuring their ability to thrive together over time
Valuation practices	**Inclusive and transparent decision-making**: Involving all members in important decisions to ensure that diverse perspectives are considered **Sharing risks and burdens**: Collective processes for distributing the need for resource intensive inputs and impacts of negative outcomes **Collaborative governance**: Developing and changing rules and norms collectively through democratic participation **Conflict resolution**: Establishing fair and low-cost processes for addressing disputes **Knowledge sharing and collective learning**: Facilitating the free exchange of information, skills, support and experiences **Monitoring and Accountability**: Developing systems to track resource use and hold members accountable for ethical behavior
Capacities	**Deliberation and facilitation**: Guiding discussions and decision-making processes to ensure equitable participation that can resolve conflict and find solutions **Social-ecological Literacy and systems thinking**: Understanding human-nature relations systems and their interconnectedness **Social-technical knowledge**: Developing skills related to maintaining or improving the community and its shared resources **Communication and organization**: Clear, respectful, and inclusive communication within the community to manage, resources, and tasks **Leadership and empathy**: Valuing the perspectives, needs, and experiences of others in the community while motivating, guiding, and inspiring others **Adaptability**: Being open to learn, change, and evolve to challenges or new information

3.4 *The Role of Valuation Practices*

Many are the "world-making practices" (Singh, 2017) of commoning. Among these, valuation practices have particular relevance for understanding the ethics of commoning (Centemeri, 2018). These are the practices through which people come to agree on what is valuable in a given situation and act accordingly so as to attain and maintain the condition deemed valuable. Valuation practices refer to processes that make actors define "the good to be pursued through engaging with the social and material environment" (Centemeri, 2018). In this sense they are enacted and implemented in the action and are not only arguments made about what is valuable and what is not (Centemeri, 2018; Euler, 2018). Beside the existence of different values, different logics might be used to determine what is valuable. These logics (often called modes of valuation) inform and are detectable in valuation practices. The logics of some valuation practices are universal and standardized, such as for example when appealing to universal ideas of justice and progress. Other valuation practices are embedded or context-dependent, and allow for the emergence of values through encounter. Rather than applying general and commensurable logics of valuation they emerge when establishing new socio-natural relationships in specific contexts, included through affects (Centemeri, 2018).

Commoning processes oriented towards social justice and environmental sustainability are characterized by so-called alternative valuation practices (Centemeri, 2018; De Angelis, 2021). These are different from dominant ones that establish a clear hierarchy of modes of valuation, in which universal and standardized goal-oriented modes of valuation are prioritized over local and emplaced modes (Centemeri, 2018; De Angelis, 2021). Commoning value practices privilege "context-dependent modes of valuation oriented by the objective of maintaining and reproducing situated human and natural life processes" (Centemeri, 2018). This does not mean that commoning does not appeal to universal modes of valuation. Valuation practices in generally act as a basis for realizing place-based connections with nature (Azzopardi et al., 2023; Berghöfer et al., 2022).

3.5 *The Role of Capacities*

Values enacted through valuation practices require that individuals and groups build capacities that enable such enactment. How can we understand the capacities individuals and collectives need to enact the values of commoning? That is, how can we understand the capacities required for ethical commoning? Capacities should empower individuals and collectives to enact their values and engage with pluralistic values. Such cultivation requires individuals and collectives to learn and build social relations to mobilize capacities (Mendis-Millard & Reed, 2007; Partelow, 2020). It requires capacities that productively bring in dialogue between people and more-than-human entities with communal relations and ethical subjectivity (Maller, 2023;

Visnic et al., 2024). As such, capacities enable a collective to reach a common goal by means that go along with the collectives' values.

Talking about capacities of individuals and groups resonates with the subjectivity-oriented literature on the commoning that emphasizes the need to pay attention to the processes and conditions necessary for "becoming in common" (Federici, 2011; Singh, 2017). To become in common, rather than seeking to fix property right issues, there is a need to prioritize the cultivation of the self as well as relations of sharing and collective practices. These capacities can be individually and collectively realized (Table 2). Individually, such as by fostering humility in the encounters with others (e.g., through listening and being attentive to their needs) while also developing courage to, for example, express, enact, or stand-up for values. But also collectively, when establishing a culture of commitment to, for example, justice that also embraces care, such as commoning processes in the permaculture movement, that redesign the subsistence sphere in accordance with principles of earth care, people care, and fair share (Centemeri, 2018), or in disaster response contexts that prioritize care and unity to overcome collective hazards (Partelow, 2020).

3.6 Values, Practices, and Capacities for a Commoning Wisdom?

There are many diverse perspectives on ethics and values in the literature. When thinking through the literature and reflecting on own research experiences, we see a strong resonance with ethical ideas such as practical wisdom and virtue ethics (Albareda & Sison, 2020; Caniglia et al., 2023). Commoners and commoning communities have strong roots in lived-experiences that provide contextual basis for enabling a person to do the right thing at the right time when facing a decision-point. To make a larger generalization, much of the commoning values, capacities and practices are aligned with ethical perspectives on practical wisdom and virtue ethics (i.e., a good person will do the right thing at the right time) (Partelow et al., 2024). We have the impression that the ethics of commoning reflect what can be referred to as 'commoning wisdom'. Commoning wisdom is the ability to do commoning well, in the contexts where it matters to the invested communities. Here we draw on virtue ethics, which suggests that an action is ethically good if the person acts and feels as they should, guided by excellent character traits (or virtues), that take into consideration situational complexity. Virtue ethics is centered around acting as part of the moral character of an individual by practicing instead of just learning rules (Russell, 2013). Such virtues are cultivated, enduring and learned by taking actions. These virtues emerge from experience, habituation and reflection. Practical wisdom is considered to be a central virtue that supports people in engaging ethically with the complexities of social and political life. Practical wisdom can be seen as the ability to make decisions and take action on practical issues that matter to people (Caniglia et al., 2023; Russell, 2013).

3.7 What is in a Framework?

When developing a new framework, Partelow (2023) notes the importance of clarifying: (1) the framework's positionality, (2) the purpose of the framework, (3) the framework's key components and relationships, (4) how it can be applied, and (5) the value the framework adds to our knowledge and applied uses. Here, we aim to briefly clarify these points for future use and engagement with our ethical commoning framework.

Positionality statements about frameworks should include: (1.a) who developed it, (1.b) the values of those researchers, (1.c) the research questions engaged with, and (1.d) the field in which it is embedded. We are located in and/or originating from North America and Western Europe. Our knowledge of commoning and the commons comes from our understanding of the published academic literature and direct experience conducting research on commons and commoning processes in Europe, Asia and the Americas, as well as experiencing the commons, cooperatives, and collective economies of our mostly urban everyday lives. We have attempted to include and reflect on a wide variety of different perspectives in developing this framework, while acknowledging shortcomings. However, our positionality may limit our understanding of more diverse perspectives and experiences with commoning, particularly from outside the Western academic institutions, especially from diverse practice communities around the world who face day-to-day realities of rights access, resources use and power struggles for livelihood security. Sustainability science is the field we most broadly associate with.

The purpose of our framework is to provide an analytical means for better understanding commoning processes and their ethics in the different contexts where they occur, with particular attention to differentiating commoning from other types of collective action or social organizational processes that may have different ethical positions.

The framework's key components are detailed in the above sections. They include the core focus of interacting values, valuation practices, and capacities. However, these components need to be considered in the context of different group structures, which broadly fell within different social-ecological contexts. These later issues are key because they give insights into to the mechanisms for collective action and institutional change, and how they may different across different group structures, and how those emergent institutions relate and are adapted to those contexts.

The framework can be applied empirically by first conceptually orienting the three major framework components to the case context. Assess how each component could be defined and measured with specific methodologies. Clear indicators may need to be developed for measurement. Develop clear research questions, and use the components and definitions in Table 2 as a guide. While the proposed framework should be modified over time based on new research and diverse inputs, using consistent and clear terms and definitions will help studies be more comparable to build theory. It would be important to distinguish the role and design of a study intending to be descriptive and observational, compared to studies that have clear normative positioning or questions. In both approaches, linking empirical insights to ethical theorizing helps improve how the field approaches building empirically informed knowledge about what ethical commoning is.

Finally, the framework adds this important dimension of comparability and reference of key terms, definitions and orientation to the literature. While the framework is not fully comprehensive and can certainly be debated, having a common object to deliberate over is important for a diverse field to discuss and debate around.

4 Some Directions Forward

Academically, in order to better understand commoning as a concept or even a theory of ethical social organization, there is a need to balance theoretical development between descriptive ethics and a sound theoretical basis which may be more normative. The interplay between both aspects is essential for having an empirically informed understanding of the ethics of commoning in the real-world. Further research needs to be able to openly and inductively observe values, practices and capacities, while also recognizing important normative features of ethics that may be generally applicable. The framework introduced above allows for describing the ethical dimensions of commoning. The framework allows openness to pursued means and ends to commoning have not been identified so far. We need to ask the question: what does it mean to do commoning and social organization well? In pursuing this question, research with the framework should facilitate investigation and comparisons of observed descriptions of commoning with normative claims of the ethics communities are presumed to want in fair, open, inclusive and democratic societies, for example. More research should strive for this balance and openness to new ideas about what ethical and effective social organization and governance processes can, do and should look like. Within this, diversity is an asset. A diversity of research, ideas and observed communities of practice will enrich debates about commoning and the tools (i.e., concepts and frameworks) we use to facilitate those debates.

References

Albareda, L., & Sison, A. J. G. (2020). Commons organizing: Embedding common good and institutions for collective action. Insights from ethics and economics. *Journal of Business Ethics, 166*(4), 727–743. https://doi.org/10.1007/s10551-020-04580-8

Anderies, J. M., & Janssen, M. A. (2016). Sustaining the commons. In *Sustaining the commons* (2.0, Vol. 2). Arizona State University. https://doi.org/10.1017/CBO9781107415324.004

Azzopardi, E., Kenter, J. O., Young, J., Leakey, C., O'Connor, S., Martino, S., Flannery, W., Sousa, L. P., Mylona, D., Frangoudes, K., Béguier, I., Pafi, M., da Silva, A. R., Ainscough, J., Koutrakis, M., da Silva, M. F., & Pita, C. (2023). What are heritage values? Integrating natural and cultural heritage into environmental valuation. *People and Nature, 5*(2), 368–383. https://doi.org/10.1002/pan3.10386

Berghöfer, U., Rode, J., Jax, K., Förster, J., Berghöfer, A., & Wittmer, H. (2022). 'Societal Relationships with Nature': A framework for understanding nature-related conflicts and multiple values. *People and Nature, 4*(2), 534–548. https://doi.org/10.1002/pan3.10305

Berta Martin-Lopez. (2020). *Plural valuation of nature matters for environmental sustainability and justice.* https://royalsociety.org/news-resources/projects/biodiversity/plural-valuation-of-nature-matters-for-environmental-sustainability-and-justice/.

Bollier, D., & Helfrich, S. (Eds.). (2015). *Patterns of commoning.* Commons Strategy Group and Off the Common Press. https://patternsofcommoning.org/

Borgatti, S. P., Mehra, a, Brass, D. J., & Labianca, G. (2009). Network analysis in the social sciences. *Science, 323*, 892–895. https://doi.org/10.1126/science.1165821

Caniglia, G., Freeth, R., Luederitz, C., Leventon, J., West, S. P., John, B., Peukert, D., Lang, D. J., von Wehrden, H., Martín-López, B., Fazey, I., Russo, F., von Wirth, T., Schlüter, M., & Vogel, C. (2023). Practical wisdom and virtue ethics for knowledge co-production in sustainability science. *Nature Sustainability.* https://doi.org/10.1038/s41893-022-01040-1

Centemeri, L. (2018). Commons and the new environmentalism of everyday life. Alternative value practices and multispecies commoning in the permaculture movement. *Rassegna Italiana Di Sociologia.* https://hal.science/hal-01773575v1

Cox, M., Arnold, G., & Tomas, S. V. (2010). A review of design principles for community-based natural resource management. *Ecology and Society, 15*(4).

Cumming, G. S. (2016). Heterarchies: Reconciling networks and hierarchies. *Trends in Ecology and Evolution, 31*(8), 622–632. https://doi.org/10.1016/j.tree.2016.04.009

De Angelis, M. (2017). *Omnia sunt communia: On the commons and the transformation to postcapitalism.* Zed Books.

De Angelis, M. (2021). *Grounding social revolution: Elements for a systems theory of commoning.* Zed Books Ltd. https://doi.org/10.5040/9781350221741.ch-006

Diprose, R. (2002). *Corporeal generosity: On giving with Nietzsche, Merleau-Ponty, and Levinas.* SUNY Press.

Euler, J. (2018). Conceptualizing the commons: Moving beyond the goods-based definition by introducing the social practices of commoning as vital determinant. *Ecological Economics, 143*, 10–16. https://doi.org/10.1016/j.ecolecon.2017.06.020

Eyster, H. N., Satterfield, T., & Chan, K. M. A. (2023). Empirical examples demonstrate how relational thinking might enrich science and practice. *People and Nature, 5*(2), 455–469. https://doi.org/10.1002/pan3.10453

Federici, S. (2011). Feminism and the politics of the commons. *The Commoner.*

Horcea-Milcu, A. I., Abson, D. J., Apetrei, C. I., Duse, I. A., Freeth, R., Riechers, M., Lam, D. P. M., Dorninger, C., & Lang, D. J. (2019). Values in transformational sustainability science: Four perspectives for change. *Sustainability Science, 14*(5), 1425–1437. https://doi.org/10.1007/s11625-019-00656-1

Kinnunen, V., & Valtonen, A. (2017). *Living ethics in a more-than-human world.* University of Lapland.

Lambek, M., Das, V., Fassin, D., & Keane, W. (2015). *Four lectures on ethics: Anthropological perspectives.* University of Chicago Press. www.haubooks.com

Maller, C. (2023). Turning things around: A discussion of values, practices, and action in the context of social-ecological change. *People and Nature, 5*(2), 258–270. https://doi.org/10.1002/pan3.10272

Maton, K. I. (2008). Empowering community settings: Agents of individual development, community betterment, and positive social change. *American Journal of Community Psychology, 41*(1–2), 4–21. https://doi.org/10.1007/s10464-007-9148-6

Mendis-Millard, S., & Reed, M. G. (2007). Understanding community capacity using adaptive and reflexive research practices: Lessons from two Canadian biosphere reserves. *Society and Natural Resources, 20*(6), 543–559. https://doi.org/10.1080/08941920601171915

Nayak, P. K., & Berkes, F. (2011). Commonisation and decommonisation: Understanding the processes of change in the Chilika Lagoon, India. *Conservation and Society, 9*(2), 132–145. https://doi.org/10.4103/0972-4923.83723

Nightingale, A. J. (2019). Commoning for inclusion? Commons, exclusion, property and socio-natural becomings. *International Journal of the Commons, 13*(1), 16. https://doi.org/10.18352/ijc.927

Ostrom, E. (1990). Governing the commons: The evolution of institutions for collective action. *Cambridge University Press.* https://doi.org/10.1017/CBO9780511807763

Partelow, S. (2020). Social capital and community disaster resilience: Post-earthquake tourism recovery on Gili Trawangan, Indonesia. *Sustainability Science, 16*, 203–220. https://doi.org/10.1007/s11625-020-00854-2

Partelow, S. (2023). What is a framework? Understanding their purpose, value, development and use. *Journal of Environmental Studies and Sciences.* https://doi.org/10.1007/s13412-023-00833-w

Partelow, S., Luederitz, C., Huang, Y.-S. (Elaine), von Wehrden, H., & Woopen, C. (2024). Building ethical awareness to strengthen co-production for transformation. *Sustainability Science.*

Partelow, S., & Manlosa, A. O. (2023). Commoning the governance: A review of literature and the integration of power. *Sustainability Science*, 0123456789. https://doi.org/10.1007/s11625-022-01191-2

Perperidis, G. (2024). A philosophical examination of the ethical foundations of the commons. *International Journal of the Commons, 18*(1), 276–287. https://doi.org/10.5334/ijc.1379

Pratson, D. F., Adams, N., & Gould, R. K. (2023). Relational values of nature in empirical research: A systematic review. In *People and nature* (Vol. 5, Issue 5, pp. 1464–1479). John Wiley and Sons Inc. https://doi.org/10.1002/pan3.10512

Pullen, A., & Rhodes, C. (2014). Corporeal ethics and the politics of resistance in organizations. *Organization, 21*(6), 782–796. https://doi.org/10.1177/1350508413484819

Russell, D. C. (2013). *The Cambridge companion to virtue ethics.* Cambridge University Press.

Singh, N. (2017). Becoming a commoner: The commons as sites for affective socio-nature encounters and co-becomings. *Ephemera: Theory & Politics in Organization, 17*(4), 751–776. www.ephemerajournal.org

van Laerhoven, F., Schoon, M., & Villamayor-Tomas, S. (2020). Celebrating the 30th anniversary of ostrom's governing the commons: Traditions and trends in the study of the commons, revisited. *International Journal of the Commons, 14*(1), 208–224. https://doi.org/10.5334/ijc.1030

Visnic, O., Maurer, M., Yoon, L., & Cook, E. M. (2024). Shifting more-than-human relationships amidst social-ecological disturbance. *People and Nature.* https://doi.org/10.1002/pan3.10687

Stefan Partelow is a professor at the Center for Life Ethics, University of Bonn. His research focuses on commons governance, collective action, and sustainability transformations in coastal and marine social–ecological systems. He combines empirical research and ethical reflection to promote just and adaptive approaches to environmental governance.

Guido Caniglia is an Associate Professor of Social Policy at the University of Helsinki, affiliated with HELSUS and the Helsinki Inequality Initiative. Holding dual PhDs in Philosophy and Biology, his interdisciplinary research spans sustainability science, epistemic justice, and transdisciplinary collaboration. He leads the ERC-funded WEIRD project, which examines sustainability through disability and queer perspectives.

Christopher Luederitz is a Professor in the Department of Mangement at the University of Quebec at Trois-Rivières. His work focuses on sustainability transformations in urban and organizational contexts, with particular attention to experimentation, living labs, and knowledge co-creation. He explores how universities, firms, and other institutions can foster systemic change for sustainability.

Open Access This chapter is licensed under the terms of the Creative Commons Attribution-NonCommercial-NoDerivatives 4.0 International License (http://creativecommons.org/licenses/by-nc-nd/4.0/), which permits any noncommercial use, sharing, distribution and reproduction in any medium or format, as long as you give appropriate credit to the original author(s) and the source, provide a link to the Creative Commons license and indicate if you modified the licensed material. You do not have permission under this license to share adapted material derived from this chapter or parts of it.

The images or other third party material in this chapter are included in the chapter's Creative Commons license, unless indicated otherwise in a credit line to the material. If material is not included in the chapter's Creative Commons license and your intended use is not permitted by statutory regulation or exceeds the permitted use, you will need to obtain permission directly from the copyright holder.

A Look at the Commons Through the Lens of Buddhist Ethics

Pisith San and Asimina Kouvara

Abstract This chapter discusses the commons through the lens of Theravada Buddhism and reflects on their relation regarding ethics and practice. First, we introduce the commons as a social system of self-organization and governance and as a way of sustainable living observed in both traditional and contemporary contexts. Simultaneously, informed by practiced examples of Buddhist commons from Cambodia, we highlight how the commons find culture-specific expressions and are compatible with the Buddhist tradition. Specifically, we discuss how Buddhist ethics align with the collaborative nature of the commons. Through the comparative understanding of both Buddhist and Commons' perspectives, we aim to contribute to the relevant literature that challenges the unsustainable global-Western paradigm of economic growth and individualism.

1 Introduction

The dominant institutions encompassing the mainstream approaches to governance and economics are premised on the Western worldview, imposed across the globe through colonization, modernization, and globalization. However, an evolving discourse highlighting the global-Western trajectory's defects calls for alternative pathways toward more sustainable and just futures. The commons play an important role in this discourse. Another notable perspective is related to Buddhism. Attention on Buddhist governance and economics relates to the re-popularized intersection of religion and governance (Ongaro & Tantardini, 2023a, 2023b), focusing on Buddhism as a world religion (Habermas, 2019) and its relevance in and for secular contexts. Additionally, the Buddhist approach is popular within the critical discourse on sustainable development, economics and politics for its potential to challenge

P. San (✉) · A. Kouvara
Ragnar Nurkse Department of Innovation and Governance, Tallinn University of Technology, Tallinn, Estonia
e-mail: pisith.san@taltech.ee; sanpisith04@gmail.com

A. Kouvara
e-mail: asimina.kouvara@taltech.ee; mina.koov@protonmail.com

© The Author(s) 2026

S. Partelow (ed.), *Ethics and the Commons*, Schriften zu Leben und Gesellschaft - Series on Life and Society 3, https://doi.org/10.1007/978-3-032-24246-4_5

established institutions (Drechsler, 2019; King, 2016; Long, 2019, 2021; Verma, 2017).

Recognizing the growing interest in these two branches of the aforementioned discourse, this chapter explores the relationship between the commons and the Buddhist approach to governance and economics, tracing similarities in their ethical and organizational aspects. We consider these similarities pivotal in discussing non-Western alternatives for more equitable and sustainable futures, transcending the individualistic and growth-oriented Western trajectory.

This chapter approaches the commons as a system of collective self-organization for managing and producing shared resources through commoning practices (Bollier & Helfrich, 2019), motivated by radically different incentives from self-interest and profit maximization. Recognizing the inherent interconnectedness of social and living systems, the commons approach is premised on ethics of care for ecological stability and the well-being of present and future generations on a planetary level. In a similar vein, key elements of Buddhist ethics, rooted in the teachings of the Buddha (c. 480–400 BCE), include interdependence, compassion, mindfulness, non-violence, and profound respect for all living beings and the environment (Harvey, 2000; Keown, 2016). Buddhism emphasizes the interdependent nature of our world and stresses the need for compassion and non-violence in our dealings with shared resources (Brown, 2019; Ng & Ng, 2020). Furthermore, mindfulness, a cornerstone of Buddhist practice, encourages conscious decision-making and an increased awareness of the consequences of our actions on the common resources. More importantly, Buddhism advocates for equity and justice, as seen in the principles of fairness and the alleviation of suffering (Jayasuriya, 2008; San et al., 2023). In this light, we posit that the commons resonates with the core Buddhist principles from an ethical perspective.

Furthermore, commons are viewed as a context-adaptive system, a pattern or template for organization and governance, that manifests in various ways within different local realities and cultural contexts (Bollier & Helfrich, 2019). These variations, however, share the same ethical foundation. There are no universal models that define exactly how ventures based on the commons should be created and operated; each example is unique (Bollier & Helfrich, 2019). Similarly, there is no blueprint on how the Buddhist canon should be implemented and institulionalised, resulting in a variety of religious customs, practices and adaptations in governance systems in Buddhist countries, according to different local and socio-historical contexts (Drechsler, 2020; Jayasuriya, 2008). Hence, from an organisational perspective both the commons and Buddhism may be viewed as context-adaptive systems that retain a solid ethical foundation. The adaptability of commons and Buddhist institutions and similarities regarding their ethical core led us to discuss their intersection as a contribution to the relevant discourse.

Additionally, although the commons as institutions of collective governance have had a significant presence throughout human history, the domination of the Western worldview has marginalized this social practice. The culture of the commons and our collective memory of it have been supplanted by the Western monoculture (Bollier, 2014), promoting individualism, imposition on nature, and a specific interpretation of progress and growth. There are cases, however, where the commons still manifest in social and everyday affairs. We argue that this holds true for Cambodia, where the Theravada Buddhist ethic continues to be fundamental to society, notwithstanding Western influence. In this direction, we focus on the Cambodian context, discussing three examples that describe how the community organizes and manages shared resources to cover spiritual and everyday needs. We approach these examples as commons-based institutions and thus refer to them as cases of 'Buddhist commons'.

The methodology involves a two-fold approach. First, we reviewed existing literature on the commons and Buddhist approaches to investigate their intersection. To delve deeper into the question of Buddhist ethics and governance, the chapter relied on the Sutta and Vinaya texts, which were initially translated from the Pali texts by different scholars. Second, we conducted semi-structured interviews with Khmer *saṅgha* members, abbots, *sālā-samnak* representatives, scholars, and experts in Khmer Buddhism and tradition in three provinces/capitals: Battambang, Siem Reap, and Phnom Penh. The selection of interviewees, comprising 20 individuals, on and off the record, was based on their involvement in Buddhist socially engaged activities and the field of Buddhist studies. During these interviews, questions were asked concerning the significant roles of "*sālā-punya/dhamma-sālā*, its foundational principles, and the decision-making process regarding the governing body and the engaged activities of *sālā-punya/dhamma-sālā*, etc.". We then employed a discourse analysis approach to analyze the interviews and derive insights into the practices and perceptions surrounding the commons practice of the Buddhist community in Cambodia. This comprehensive methodology enabled us to better understand how Buddhist commons function in the Cambodian case, gaining insights for the broader discussion on alternative pathways of governance and economics focusing on the commons' potential.

2 The Buddhist Ethics and the Commons

Buddhism began on the Indian subcontinent in the fifth century BC as a small religious community that developed a certain distance, both self-perceived and realization, from other contemporary religious communities of the time (Reynolds & Hallisey, 1987). From the earliest date, there were signs of an emerging commons practice in the Buddhist community regarding the decision-making process, governance, and the recruitment of new members into the *saṅgha* community through open participation.

The main goal of joining the *saṅgha* community is to liberate oneself from worldly attachment and to pursue awakening by living a dedicated life of spiritual practice. However, this goal often leads to its characterization as an 'otherworldly' religion with a gnostic aversion to worldly matters, sparking debates about its level of social engagement (Harris, 1999). In Weber's 1916 analysis of the comparative sociology of Indian religions, he denied the relationship between early Indian Buddhist traditions and social-economic engagement. Weber portrayed the belief systems in India, including Buddhism, as 'otherworldly' religions that emphasize individual salvation over social responsibility.

The view of Buddhism as 'otherworldly' leaves its social role, economic engagement, and organization largely unexplored (Atwood, 1996). In theory and practice, Buddhism is widely seen as an essential source of non-profit activity and social welfare. The end goal of Buddhism is, of course, to achieve *nirvana,* a state of liberation from worldly attachment, however, the *saṅgha* community does not entirely detach themselves from worldly matters. The principles, practice, and governance of the Buddhist community especially the *saṅgha* are rooted in open participation, shared purposes and resources, mutuality, and fairness which can be viewed as a sign of social engagement or the commons.

It can be assumed that the Buddha views the ways of living and governing society are different from Machiavelli (1469–1527), an Italian philosopher, writer, and politician, who emphasizes the importance of identifying one's other weaknesses and vulnerabilities and how one can exploit them and use them to one's advantage. In *The Prince*, Machiavelli (1532, p. 91) warns that evil prevails even when someone wishes to adhere solely to virtuous principles.

In a similar argument, Smith (1723–1790), a Scottish philosopher and economist, asserts that human action and decisions are driven by self-interest because most people strive to maximize their happiness, well-being, or success, even at the expense of others. According to Smith, the individualistic nature of humans fosters competition, motivating producers to reduce prices and improve the quality of their products by increasing efficiency and innovation, contributing to economic growth; thus benefiting all members of society, including the poorest.

Like Smith, Hardin (1915–2003), an American ecologist, argues that humans are inherently self-interested, and that this self-interest drives much of their behavior. Particularly, in situations where a resource is held in common and there are no clear rules or regulations for its use, self-interest urges people to exploit the resource until it is exhausted. In *The Tragedy of the Commons*, Hardin (1968, p. 1244) argues that individuals benefit personally from their ability to ignore the truth, even though it may harm society as a whole. Ultimately, this understanding of behavior results in the collapse and tragedy of the commons. As Hardin suggests, a solution to this problem is the enclosure of resources through government regulation and private ownership. Enclosure, he argues, could prevent the depletion and overuse of resources by creating individual incentives to conserve them.

Premised upon such assumptions is the dominant global-Western trajectory, according to which economic growth is a prerequisite for the social prosperity (Pansera & Fressoli, 2021). Typical mechanisms to achieve economic growth include the enclosure, privatization, commercialization, and central governance of common resources; efficiency-oriented industrial production; and unilateral technological progress. These mechanisms prioritize monetary profit maximization and have profound consequences related to social inequalities, the environmental crisis, and the extinction of cultural diversity. Despite the acknowledged defects currently posing an existential threat to humans and non-humans alike, the Western trajectory continues to prevail as the only alternative. Nevertheless, adherence to the homogenizing and fragmentary logic of this trajectory has long been a subject of controversy.

A notable critique towards the mainstream perspectives is Schumacher's work on Buddhist Economics, which gained widespread attention following the publication of his book *Small is Beautiful: Economics as if People Mattered* in 1973. The book presented Schumacher's alternative economic theories that critiqued mainstream Euro-American economic thinking and proposed alternative development models for Asia and the West. Schumacher's theories emphasize human-scale technologies, decentralization, and the sustainable use of natural resources aligned with the Buddhist perspective, shedding light onto the social, political and economic aspects of Buddhism. As opposed to Adam Smith, who emphasizes the role of self-interest in promoting economic growth, Schumacher argues that pursuing self-interest in a free-market system might lead to exploiting workers and the environment. Rather than maximizing profit, Schumacher claims that economic development should not be driven by self-interest alone but instead be based on ethical values and sustainable principles to serve the needs of the people.

Schumacher's self-interest argument is well aligned with the concept of *anattā*, meaning non-self in Buddhism, referring to the concept that no permanent, unchanging self or soul is found in any human being. In Buddhism, all phenomena, including individuals, are impermanent and constantly changing, and none possess an inherent or intrinsic self-nature. The *anattā* principle aims to encourage Buddhist practitioners to detach themselves from the misguided clinging to what is mistakenly considered to be *self*, and through that detachment (alongside moral livelihood and meditation), the path to *nirvana*—a state of liberation from the cycle of suffering—can be successfully traversed.

From a different viewpoint, Payutto (1992, p. 25), a well-known scholar in Buddhist Economics, views self-interest as a natural aspect of human behavior, and thus the driving force of competition in mainstream economics. However, Payutto emphasizes that competition must not be pursued at the expense of others or environmental degradation. According to Payutto, such competition is driven by *taṇhā* (sensual desire), considered a leading cause of suffering and an unwholesome state in Buddhism. Therefore, Payutto (1992, p. 57) stresses that human behavior is also driven by *chanda*, or willingness, regarded as a wholesome state for bringing about real well-being or quality of life. He continues to address that all individuals have a moral responsibility to act in a way that does not harm others or the environment; since non-harming and the idea that moral actions have consequences (*karma*) over human existence are essential aspects of Buddhist teachings (Keown, 1996; Payutto, 1994).

Based on the doctrine of *karma*, one of the central teachings in Buddhism, every action has consequences (Harvey, 2000; Rahula, 1959). According to Buddhism, *karma* is not just an inheritance from a past life that determines and dictates one's fate. Buddhism points out that *karma* originates from the threefold nature of one's daily activities: bodily action, verbal action, and mental action, accompanied by the pulling feeling of volition (Ghose, 2007). Therefore, Buddhism emphasizes that moral actions have consequences, affecting individuals and the states of affairs brought into being through moral acts (Keown, 1996).

Rooted in the theory of *paṭiccasamuppāda*, or the theory of Indra's Jewel Net in Mahayana Buddhism, more precisely translated as "dependent origination" or "dependent co-arising," all things arise in dependence upon other things,[1] meaning that nothing can stand on its own (see Payutto, 1994; Keefe, 1997; Thanissaro, 2013a). According to this theory, Buddhism views the world as a vast interconnected web of events, where each phenomenon constitutes and reflects other phenomena. The *paṭiccasamuppāda* theory or the theory of Indra's Jewel Net provides an idea with powerful ethical and political implications: "If we are all part of a vast, interdependent network of being, what we do can have profound effects on others as our actions reverberate throughout this network" (McMahan, 2008, p. 132). The doctrine demonstrates that every action has its consequences. These consequences first affect the individual who performed the action and then extend to the people around them and society as a whole. The interconnected nature of reality, means that '[o]ur ethics—and the behavior that naturally flows from our ethics—contribute to the causes and conditions that determine who we are, the kind of society we live in and the condition of our environment' (Payutto, 1994). To understand connectedness is the key to imagine a different vision for politics (Long, 2021).

Known for her research on collective action and common-pool resources (CPRs), Ostrom challenges the conventional view that individuals are always motivated by self-interest and that common-pool resources (such as fisheries, forests, and irrigation systems) are destined to be overused and depleted without centralized government regulation or privatization. Ostrom argues that Hardin's perspective is oversimplified and fails to consider the complexity of human behavior and social systems. Despite potential conflicts of interest and the possibility for opportunistic behaviors to arise within commons-based institutions, Ostrom's work, informed by hundreds of cases from around the world, asserts that individuals can act in the interest of the group or community, self-organize their own systems of commons-based governance, and develop a cooperative ethic to manage shared resources and achieve shared goals (Ostrom, 2015). Similarly, Benkler (2006, 2011), referring to cases of commons-based peer-production (CBPP), such as Wikipedia, or Free/Libre and open-source-software (FLOSS), explains that people have a natural inclination towards collaboration, which can counteract individualistic and self-interested behavior. Evolutionary scientists also support this tendency (Bollier, 2014).

To address the ambiguity of the term (Hess, 2008), this chapter defines the commons as a social system whereby a community self-organizes and co-devises protocols, values, and norms to manage and maintain shared goods and resources; or collectively produce goods and services. These encompass natural resources and a wide array of tangible and intangible human creations (e.g., data, software, archives, techniques, knowledge, and cultural heritage) (Bauwens et al., 2019; Bollier & Helfrich, 2019; Hess, 2008; Lekakis, 2020). The commons ensure access to these

[1] See *Vera Sutta: Animosity (AN 10.92),* translated from the Pali text by Thanissaro (2013a), available online at www.accesstoinsight.org

resources, goods and services, and to infrastructures and decisions. This approach stands in stark contrast to exclusive control, whether by private entities or the state, where access can be potentially excluded (particularly considering that the presence of influential economic interests in the public sphere further complicates the idea of public ownership).

The commons is framed by values radically different from those of the market economy, promoting ethics of sufficiency economy, stewardship and knowledge sovereignty (Bollier, 2014; Bollier & Helfrich, 2019; Kostakis & Pazaitis, 2020). Beyond monetary thinking and exchange value, trust, reciprocity, and relationality are put forward (Mandalaki & Fotaki, 2020). At its core, the commons logic recognizes our corporeal vulnerability, and our dependence on each other and nature; while stressing the need to acknowledge that building our lives upon the destruction of ecosystems, the exploitation of resources, and the suffering of others, will only keep things on the same damaging track they currently are (Bollier & Helfrich, 2019; Mandalaki & Fotaki, 2020). Put differently, the commons emphasizes interconnectedness, and calls for taking responsibility of our actions. As Bollier (2014) explains, in its deepest reaches, the commons extends beyond economics, public policy, or politics, indicating a distinct mode of human existence (ontology) compared to those we are accustomed to.

Despite being characterized as an 'otherworldly' religion with a gnostic aversion to worldly matters, sparking debates about its level of social engagement, emphasizing individual salvation over social responsibility (Harris, 1999; Weber, 1958), according to mainstream approaches, Buddhism is primarily discussed as a set of ethical guidelines on how to lead one's life towards liberation and the commons as a mode of organization and governance (Mandalaki & Fotaki, 2020), both expand to ontological and political spheres. There are no rigid rules on how to implement and institutionalize the Buddhist canon, nor a blueprint of how commons ventures should emerge. Both cases, however, are strongly tied to a solid ethical core that emphasizes the notion of interdependence point towards justice, sustainability, and human flourishing. The soft adaptable character of commons- and Buddhism-based institutions anchored in similar ethics for socio-ecological prosperity, has allowed a variety of renderings and ventures to emerge in compliance with local and cultural contexts, community needs, and visions.

Next, we explore three examples of institutions from Cambodia, where Theravada Buddhism is integral to society and everyday life. These examples describe how the religious community self-organizes and manages shared resources to cover spiritual and everyday needs. On one hand, these examples of Buddhist institutions comprise a context-specific implementation of the Buddhist canon. On the other hand, they demonstrate a way of governance that resonates with the organizational aspect of the commons. In this light, we look into these institutions through the lens of the commons, understanding them as secular, context-specific adaptations of commons-based organization, and thus refer to them as 'Buddhist commons'.

3 The *Saṅgha* Community

Among the three cases of the commons that we discuss here, the *saṅgha* community is one of the significant cases. The reason that the *saṅgha* community is relevant here is that the element of the commons emerges in the daily practice and the community governance. The *saṅgha* is a monastic community and fundamental institution in Buddhism, comprising male-ordained and female-ordained communities, who renounced worldly attachments and committed themselves to the pursuit of spiritual awakening by living a dedicated life of spiritual practice, study, and service. The term '*saṅgha*' is a Pali term meaning an 'assembly,' 'association,' 'community,' or 'order' and is most commonly used to refer to an order of Buddhist monks or nuns (De, 1970; Buswell & Lopez, 2014; Borchert, 2017).

The *saṅgha* community is governed by a set of monastic rules, which provide ethical guidelines for their daily practice, and create a space for the commons to emerge. In principle, the *upajjhāya*, spiritual preceptors, occupy the highest position in the *saṅgha* community, however, there is no absolute power over the governing or the decision-making process in the *saṅgha* community. The decision-making process is open to all *saṅgha* members regardless of their seniority or status. For instance, in the recruitment of any new members, the *saṅgha* community embraces the bottom-up approach that is contrary to top-down or authoritarian approaches; it therefore works toward inclusivity through consensus decision-making (Dutt, 1924; Jinananda, 1961; Heng, 2022; San, 2024; San & Shakya, 2025; Prebish, 2018).

In the process of joining the *saṅgha* community, the candidate seeks the approval of the chief of the *saṅgha* community. The approval needs to be done through the collective decision, requiring no fewer than ten *saṅgha* members, except in bordering regions, where the monastic rules allow the higher ordination to be held in the meeting of four *saṅgha* members (Dutt, 1924, p. 147; Dickson, 1963, p. 14). Before deciding to accept the candidate into the *saṅgha* community, two mentors are appointed by the spiritual preceptor to assess the candidate's background. Once the investigation is completed, the mentors verbally propose the candidate's ordination three times to the assembly of *saṅgha*. In the absence of objections, silence is expected from all present. However, if an objection arises, it necessitates a vocal expression, and the process will be repeated until a consensus is reached. Likewise, appointing individuals to positions of responsibility, such as inventory manager, requires the endorsement of the members residing in the temple. Moreover, these appointments are bound by a stringent legal process, as stipulated in the sacred scriptures (Heng, 1998, p. 10).

If one *saṅgha* member is very sick and unable to join the meeting in person, he must remain outside the boundary of the monastery, or he may send his consent through the other, which is called *chanda* in Pali term, as a sign of pre-agreement with the decision made by the monastic community. Any decision made without even one monk's presence is invalid (Dutt, 1924, p. 146). In response to the severity of a given issue in the *saṅgha* community, the decision-making process needs to be held through the proper performance of *saṅgha-kamma* (the *saṅgha's* formal act). *Saṅgha-kamma* is employed for various purposes, such as reaching agreements, making decisions, or taking actions within the *saṅgha* assembly, which comprises the following requisites (Dutt, 1924, p. 125):

1. The presence of the proper number of competent *saṅgha* members
2. The conveyance of all absentee ballots
3. The motion being proposed
4. The proper proclamation of the proposed act

Monastic life is set up as a ruling system in which leaders are chosen by their qualities and with the approval of the *saṅgha* assembly. Each *saṅgha* member is required to participate in maintaining the stability of the rule of law of the monastery, i.e., participating every two weeks in a ceremony known as *uposatha* (bi-weekly meeting) in the monastery to review compliance with the monastic code. In the Mahāparinibbāna Sutta (DN 16—Thanisssaro, 2013b), the Buddha lists seven conditions that will maintain the unity and solidarity of the *saṅgha* community. The first two are these: "(1) As long as the bhikkhus meet often, meet a great deal, their growth can be expected, not their decline. (2) As long as the bhikkhus meet in unity, adjourn from their meetings in unity, and conduct Community business in unity, their growth can be expected, not their decline" (DN 16—Thanissaro, 2013b).

The *uposatha* observance was formulated to fulfill these purposes, serving as a bi-weekly opportunity for the *saṅgha* assembly to gather, update their membership rolls, address issues, and reaffirm their common adherence to the monastic code (Thanissaro, 2013c, p. 1098). This practice is also observed in the Ostrom Design Principles, particularly in the collective-choice arrangements to ensure that those affected by the rules can participate in modifying them (Ostrom, 2008, p. 224). Performing *uposatha* with an incomplete or divided *saṅgha* assembly is regarded as wrong-doing according to the monastic code. Therefore, the first duty is to convey consent and purity on behalf of a *saṅgha* member who cannot attend the *uposatha* observance. In the Mahāvagga,[2] the Buddha instructed that a sick *saṅgha* member, unable to participate in the *uposatha*, should communicate his consent and purity through another *saṅgha* member to those attending the *uposatha.* According to the monastic code,[3] any *saṅgha* member who witnesses another monk's transgression but fails to report it to the *saṅgha* assembly is also considered to be committing an offense (Rhys-Davids & Oldenberg, 1881, p. 33; Ñāṇatusita, 2014, p. 174). Monitoring and sanctioning, as Ostrom (2008, p. 224) points out, are core principles for communities to sustainably and equitably govern their commons.

Overall, the *saṅgha* in Buddhism lives together as a community, pursuing liberation by adhering to the rules and regulations set forth by the Buddha to maintain continuity and unity within the *saṅgha* assembly. Unity, purity, and integrity are regarded as key factors in preserving the *saṅgha* institution. Democracy or a consensus-based approach is used to make decisions, reach agreements, and act within the *saṅgha* assembly through *saṅgha-kamma* performances. While the primary purpose

---

[2] Maḥāvagga, Khandhaka II, Chapter 22, translated from the Pali text by Rhys-Davids & Oldenberg (1881, pp. 274-5).

[3] In the Suddhika-pācittiya: Requiring of a Transgression for Purification.

of joining the *saṅgha* is to liberate oneself from worldly attachments, the *saṅgha* typically remains connected to lay society because their daily basic needs rely on it, and they also have duties as the *dharma* messengers, to disseminate and teach the principles of *dhamma* to lay society.

3.1 *The Buddhist Temple (*Wat*)*

In Khmer Buddhist community, the term *wat* (see Fig. 1) refers to a monastery, pagoda, or temple. A *wat* serves as a residence for the *saṅgha* members and, at the same time, also functions as a center for spiritual practice, cultural heritage, and community activities. If we were to ask whether a Buddhist temple belongs to someone, undoubtedly, no one would dare claim ownership of a Buddhist temple in Cambodia. According to a report released by the Ministry of Cults and Religions in 2022, there are 5104 Buddhist temples in Cambodia (FreshNews, 2022). Of the 5104 temples, some are newly established, while others proudly bear a history of four to five centuries.

Buddhist Temple is regarded as common property, generally recognized as collective assets established and maintained by the Buddhist community under the guidance of the *saṅgha* community and the temple committee. Their primary purpose is to serve the common good within their respective communities. This characteristic may account for the prevailing sentiment among the public that the temple is akin to an integral part of their home and enhances a sense of community belonging, even though they do not reside there directly.

Fig. 1 Roluos Temple in Roluos Commune, Prasat Bakong District, Siem Reap Province, Cambodia. *Source* Author. The image depicts a Cambodian Buddhist temple along a street with trees and light poles

The *saṅgha* community relies on donations from the laity. Alongside meditation and virtuous practices, donating land or other material necessities to the *saṅgha* community is an act of accumulating merit in Theravada Buddhism. Furthermore, donation is also viewed as a form of renunciation within the Buddhist belief, signifying a virtuous act—a means to sever one's attachments to worldly possessions and to be free from greed (Falk, 2007, p. 140). When understood in this light, *saṅgha* members are not expected to reciprocate in any manner. Tambiah (1970, p. 213) referred to this as 'a double negation of reciprocity': on the one hand, the donor can liberate themselves from worldly attachments, and on the other, the recipient is not obliged to 'repay' following the customary logic of gift exchange, as described by Mauss ([1925] 1990), or adhere to the *quid pro quo* principle found in commercial transactions.

Building upon Tambiah's concept of the 'double negation of reciprocity' (1970, p. 213), Strenski (1983) elaborates that as the *saṅgha* community receives productive lands and durable items, it manages these assets to 'enrich society at large.' Consequently, the *saṅgha* community doesn't directly reciprocate the donors but indirectly benefits a third party, eventually returning benefits to the original donor (Strenski, 1983, p. 473). Strenski's 'circle of giving' concept provides a lens for understanding how Buddhist temples function as a commons.

All donations received by the *saṅgha* community from the laity are considered common property, meant to be shared and accessible to all members of the community. These objects are known as *garubhandha,* meaning 'heavy objects' or 'expensive goods.' This pedagogical term encompasses a diverse array of equipment and donations generously offered by lay followers to the *saṅgha* community in the temple. The concept of 'heavy objects' goes beyond their mere physical weight; it also signifies the substantial responsibility of effectively managing and judiciously allocating these communal resources, embodying principles of equity and justice.

Garubhandha cannot be owned by any individual *saṅgha* member or given from one *saṅgha* member to another. The *garubhandha* property is designated for the *saṅgha* community, reflecting its collective ownership and responsibility. However, the *saṅgha* community may choose to temporarily loan *garubhandha* to others, ensuring its utilization benefits the wider community's welfare and needs. Through this thoughtful management of shared assets, the *saṅgha* community upholds the values of compassion and interconnectedness, reflecting the teachings of the Buddha.

The Buddha enacted many relevant laws to avoid exploiting common property to serve personal interests. For instance, the *saṅgha* members are prohibited from treating common property as their personal property. Instead, if a monk requires the use of such belongings, they may borrow or exchange them for items of equal value with the consent of the *saṅgha* community. Moreover, any *saṅgha* member who dares to hand over this common property to another individual, be it a fellow *saṅgha* member or a devotee of the temple, without the consent of the *saṅgha* community will be fined as prescribed by the rules and regulations outlined in the books of discipline.

3.2 The Sālā-Samnak

Beyond the *saṅgha* community, other expressions of the commons are observed within the lay Buddhist community. *Sālā-samnak* is a gathering place for lay Buddhists to engage in various activities, such as doing charity, community meetings, hosting festivals, and organizing ceremonies. *Sālā-samnak* means a rest house in Cambodian, traditionally built by villagers on the roadside as a rest stop for travelers and passersby. *Sālā-samnak* is normally found on either side of the road but never far from the village.

The construction of *sālā-samnak* is an old tradition in Cambodia. This tradition became even more important when King Jayavarma VII (c. 1122–1218) came to power. According to one of our interviewees, Ang Choulean, a Khmer anthropologist and a professor of historical anthropology at the Royal University of Fine Arts based in Phnom Penh, the idea of building a resting house in the Khmer Empire might have existed even before the reign of King Jayavarma VII; however, its implementation was scattered and unsystematic. King Jayavarma VII was the one who put it into state policy and made it more systematic based on the principles of Buddhism. From the verses 122–126 of the inscriptions on Preah Khan stele dated from 1191 CE, we learn that the King built 121 *sālā-samnaks* on the main roads leading from the Angkor capital, Yasodharapura, to distant areas. There were 57 rest houses with fire as staging posts along the road from Yasodharapura to the city of Campā, which is now located in Vietnam; and 17 rest houses along the road from Yasodharapura to Vimāyapura, which is now located in north-east Thailand (Maxwell, 2007, pp. 84–85).

Finot (1925, pp. 421–422) used another Sanskrit word, 'dharmasālā,' to interpret these structures since he considered the highways as pilgrimage routes and the buildings beside them as religious hostels. He noted that the reason for considering them *dharmasālās* is the presence of Lokeśvara Bodhisattva, which offers protection against dangers such as brigands, elephants, snakes, and wild beasts. Although the term 'dharmasālā' doesn't appear in the Preah Khan inscription, it has since become widely used to refer to these rest houses. In a first-hand account of Khmer civilization written by a Chinese envoy who resided in Angkor for a year between 1296–1297, the Khmer referred to these resting places as *'sen-mu'* (Khmer, *samnak*) (Chou, 1992, p. 65).

In the practice of performing generosity, providing a free *sālā-samnak* to travelers and passersby is an expression of collective hospitality and generosity. More importantly, this is a grassroots initiative by the people to give back to the community. In some cases, *sālā-samnak* is interchangeably called *sālā-bun* (a merit-making hall) for a slightly additional function and purpose.

The main purpose of *sālā-bun* (see Fig. 2) is to offer a nearby location for the elderly members to conduct meritorious activities in their Buddhist communities. Apart from merit-making activities, *sālā-bun* serves as a community gathering place for traditional festivals, especially for the 'village festival' that takes place after the harvest. Although *sālā-bun* is primarily a Buddhist ceremonial place, it also offers a free resting place for travelers and passersby. Due to this, *sālā-bun* is sometimes referred to as *sālā-chortean*, which means the hall where six items are donated. The *sālā-chortean* contains six items that travelers and passerby can use for free: a sleeping place, a water jar, a toilet, a mosquito net with mattress, pillows, and traditional medicine.

In most cases, *sālā-buns* are built for those who live far from Buddhist temples to make it easier for them to visit and perform meritorious deeds in their community. In the community, whenever there is a Buddhist festival, the monks in the nearby pagodas are always invited to participate. Building *sālā-bun* in the communities has made it easier for the elderly, most of whom find it difficult to travel to the temples.

We interviewed Ian Oeun, 76, who donated his land to build a *sālā-bun* in Moung Russey District, Battambang Province, Cambodia. He explained that since the Buddhist temple was too far from where he lived, making it difficult for the elderly to perform traditional Buddhist ceremonies, he and the people in his community decided to build a *sālā-bun*. He explained that he was inspired to donate his land for the benefit of his community after listening to a Buddhist monk preaching about charitable acts; particularly the Jataka story of a man named Magha performing public services by constructing roads and rest houses for the public. Oeun emphasizes that even though he is the land donor for the *sālā-bun*, all community members financially contributed to its construction. Therefore, he states that this *sālā-bun* is not exclusively his property but belongs to the community. Community members have an equal right to manage and use it according to their needs.

Fig. 2 Sālā-bun in Wat Chork Community, Kakaoh Commune, Moung Ruessei District, Battambang Province, Cambodia. *Source* Author. The image depicts a Cambodian resting house along road with a bicycle outside

According to one of our interviewees, Venerable Nhory San, the abbot of Wat Chork Thom monastery in Moung Ruessei District, Battambang Province, Cambodia, the *sālā-bun* has played a significant role in serving the public interest in the Buddhist community. He explains that the *sālā-bun* has made it possible for the elderly to perform charitable activities without traveling long distances to the pagoda. He also noted that the role of the *sālā-bun* has changed significantly over time, with fewer travelers and passersby nowadays. However, it has become a place for offering alms and performing meritorious deeds following Buddhism.

In other communities, the *sālā-bun* serves not only as a place for Buddhist ceremonies but also as a means to reduce poverty. In our interview with Tung Da, a representative of the *sālā-bun* in the Chork Thom community in Moung Russey district, he told us that the *sālā-bun* in his community had established a fund to support the elderly, including paying for medical treatment and providing loans at a low-interest rate to its members. Most of the fund's income comes from renting community equipment at the lowest possible prices and partly from fundraising. All funds derived from equipment leasing and fundraising are managed directly by community members without interference from public or private institutions.

4 Discussion

The relationship between religion and society has long been a subject of profound interest and significance, as noted by scholars like Ongaro and Tantardini (2023a, 2023b). The influence of religion in the public realm has shaped societies, norms, and power structures throughout history. One particular religious tradition that has gathered significant attention is Buddhism. Its uniqueness as a world religion, as highlighted by Habermas (Foshay, 2009; Habermas, 2019; San et al., 2023), extends beyond the spiritual and philosophical aspects.

Buddhism refers to how one should lead their life, avoiding causing harm to oneself, those around them, and society at large. The fundamental principles of interconnectedness, compassion, moderation, and simplicity, believed to be the pathway towards a rightful and sustainable life, are incorporated into the public sphere and people's daily lives. Hence, this ethical approach also reflects in the ways the Buddhist community organizes, makes decisions and manages resources.

Informed by the case of Cambodia, this interaction between religion and society is manifested through the relationship between the *saṅgha* and the lay community in their daily practices. The *wat* and *sālā-samnak* serve as platforms to facilitate these interactions. The *saṅgha* as an institution and its symbiotic relationship with the laity, the *wat* as a communal space dedicated to spiritual activities, and the *sālā-samnak* as a community space for secular activities, are primarily purposed to serve the common good anchored in Buddhist ethics.

The three cases of the commons practices in the Cambodian Buddhist community demonstrate how Buddhist ethics, values, and belief systems play a significant role in driving our society which is working beyond the market-oriented economy. What makes these three cases unique is that they run their institutions both spiritually and secularly, based on the principles deeply rooted in Buddhist ethics. Motivated by the pursuit of individual and collective good, they are driven by the concept of making merit rather than self-interest.

As discussed earlier, the principles employed by the Buddhist community to govern their institutions often reflect elements of the commons, closely aligning with Ostrom's principles such as monitoring, sanctioning, shared rights, voting, consensus-based decision-making, and collective choice rules. These principles have been demonstrated as key factors within the Buddhist community, shaping their motivations for collective behavior and enabling them to govern their commons in an ethical and sustainable way.

From a different perspective, the aforementioned examples demonstrate how religion and spirituality manifest within the public realm as a commons, given that equity, inclusion, reciprocity and sharing are essential to their operation and sustenance. Furthermore, considering the ontological dimension of the commons as a mode of human existence that is mindful of individual, collective and planetary well-being, the emergence of Buddhist commons is a natural outcome. Put differently, Buddhist commons illustrate how Buddhism and the commons align, both from an organizational and ontological point of view, premised upon a mutual ethical framework and purpose.

The discussion on Buddhist commons is an opportunity to challenge the hegemony of the Western worldview; as it is boldly expressed through mainstream economics and the cultivation of a global monoculture promoting individualistic and antagonistic ways of living. Contrastingly, Buddhist commons demonstrates the dynamic character of the commons as a living system that can adapt in diverse contexts, not tied to a unilateral one-size-fits-all model. Moreover, shows an effective and timeless alternative of a socio-economic paradigm in practice.

Despite living in a society dominated by the market economy, Buddhist communities are still able to organize their institutions in a commons-based manner, framed by principles that transcend individualistic self-interest. These principles have served the needs of local people and kept their institutions functioning for centuries. Therefore, assuming that prerequisite of the social well-being is a growth-oriented economy driven by self-interest, profit-maximization and resource exploitation, in accordance with the global-Western trajectory, despite its deep consequences, is debatable.

On the contrary, the core idea of interconnectedness, interdependence, or "interbeing" (Hanh, 1991), as it pertains to both Buddhism and the commons, addresses the vulnerability of human existence and how much we depend on each other and nature. As pointed out in the theory of *paṭiccasamuppāda*, i.e., "dependent origination" or "dependent co-arising", all things arise in dependence upon other things. The idea of interbeing emphasizes that building our lives on ecosystem destruction and the suffering of others will only cause further suffering. But, also brings forward the potential of collective action and compassion towards all things for leading a healthy life, serving the common good and building sustainable futures.

5 Conclusions

In this chapter, we explored the relationship between Buddhism and the commons, informed by examples of the practices of the Buddhist *saṅgha* and lay community in Cambodia. The inquiry delved into the influence of Buddhism in public affairs, despite the common perception of it as an apolitical tradition (Weber, 1988). In this direction, we explored its influence on society, governance mechanisms, and resource management, looking beyond its spiritual and philosophical dimensions. Additionally, we underscored the ontological dimension of the commons, transcending the realms of politics, economics, and public policy.

Through the discussion on Buddhist commons, we emphasized the contextuality and adaptability of the commons as a social system of governance and co-management of resources. Furthermore, we pointed out that its ethical foundations are grounded in the ethos of interconnectedness, aiming for human and non-human well-being on individual, collective and planetary levels. In this sense, we posed that the commons is compatible with the fundamental Buddhist principles, and, hence, has naturally emerged within the Buddhist society.

In this chapter, we tried to demonstrate that Buddhism strongly emphasizes that everything is interconnected; nothing can stand alone. Therefore, it is very important to lead a life by avoiding causing harm to those around us, whether they are human beings, nature, or society at large. The compassion, moderation, and simplicity embedded in Buddhist ethics, believed to be the pathway towards a rightful and sustainable life, are incorporated into the public sphere and people's daily lives. This ethical approach reflects in the ways the Buddhist community organizes, makes decisions, and manages resources, which, as we argued earlier, aligns well with the principles of the commons.

Overall, Buddhist commons exemplify the conjunction between the spiritual and the secular in the public realm. Moreover, they provide a unique and practiced alternative of a socio-economic trajectory, challenging the ontological foundations of the dominant global-Western paradigm. Ultimately, Buddhist commons demonstrate the timeless potential of the commons to foster social and planetary harmony.

Funding This book chapter is part of a project, COSMOLOCALISM, that has received funding from the European Research Council (ERC) under the European Union's Horizon 2020 research and innovation programme (grant agreement No. 802512).

Bibliography

Atwood, C. P. (1996). Buddhism and popular ritual in Mongolian religion: A reexamination of the fire cult. *History of Religions, 36*(2), 112–139.

Bauwens, M., Kostakis, V., & Pazaitis, A. (2019). *Peer to peer: The commons manifesto.* University of Westminster Press. https://doi.org/10.16997/book33

Benkler, Y. (2006). *The wealth of networks: How social production transforms markets and freedom.* Yale University Press.

Benkler, Y. (2011). *The Penguin and the Leviathan: How cooperation triumphs over self-interest.* Crown Business.

Bollier, D. (2014). *Think like a commoner: A short introduction to the life of the commons.* New Society Publishers.

Bollier, D., & Helfrich, S. (2019). *Free, fair, and alive: The insurgent power of the commons.* New Society Publishers.

Borchert, T. (2017). The Sangha as an institution. In *Oxford research encyclopedia of religion.* https://doi.org/10.1093/acrefore/9780199340378.013.194

Brown, C. (2019). *Buddhist economics: An enlightened approach to the dismal science.* Bloomsbury Press.

Buswell, R. E., & Lopez, D. S. (2014). *The Princeton dictionary of Buddhism.* Princeton University Press.

Chou, T.-K. (1992). *The customs of Cambodia* (2nd ed.). Siam Society.

Choulean, A. (2017). *Sala-samnak.* Retrieved April 2023, from Yosothor: https://www.yosothor.org/publications/khmer-renaissance/chapter-four/sala-somnak.html

De L. Young, D. N. (1970). The Sangha in Buddhist history. *Religious Studies, 6*(3), 243–252.

Dickson, J. F. (1963). *Ordination in Theravada Buddhism* (T. Piyadassi, Ed.). Buddhist Publication Society.

Drechsler, W. (2019). The reality and diversity of Buddhist economics. *American Journal of Economics and Sociology, 78*(2), 524–560.

Drechsler, W. (2020). Buddhist economics for pandemic times: A review of Buddhist economics. An enlightened approach to the dismal science. *Journal of Global Buddhism, 21*, 103–109.

Dutt, S. (1924). *Early Buddhist Monachism: 600 BC-100 BC.* Kegan Paul, Trench, Trubner & Co., Ltd.

Falk, M. L. (2007). *Making fields of merit: Buddhist female ascetics and gendered orders in Thailand.* NIAS Press.

Finot, L. (1925). Dharmaçâlâs au Cambodge. *École française d'Extrême-Orient, 25*(3/4), 417–422.

Foshay, R. (2009). Tension on the left: Buddhist ethics and postmodernism in Habermas and Wilber. *Journal of Integral Theory & Practice, 4*(3), 127–140. https://www.academia.edu/3478736/Tension_on_the_Left_Buddhist_Ethics_and_Postmodernism_in_Habermas_and_Wilber

FreshNews. (2022). *Spokesperson of the ministry of cults and religion: Cambodia currently has 5,104 pagodas and more than 70,000 monks.* Retrieved November 2023, from Fresh News: https://freshnewsasia.com/index.php/en/localnews/259006-2022-09-29-11-25-22.html

Ghose, L. (2007). Karma and the possibility of purification: An ethical and psychological analysis of the doctrine of karma in Buddhism. *Journal of Religious Ethics, 35*(2), 259–290. https://doi.org/10.1111/j.1467-9795.2007.00306.x

Hanh, T. N. (1991). *Peace is every step: The path of mindfulness in everyday life* (A. Kotler, Ed.). Bantam.

Habermas, J. (2019). *Auch eine Geschichte der Philosophie 1: Die okzidentale Konstellation von Glauben und Wissen.* Suhrkamp.

Hardin, G. (1968). The tragedy of the commons. *Sciences, 162*, 1243–1248.

Harris, I. (1999). Buddhism in extremis: The case of Cambodia. *Buddhism and Politics in Twentieth-Century Asia, London and New York: Continuum.* 54–78.

Harvey, P. (2000). *An introduction to Buddhist ethics: Foundations.* Cambridge University Press.

Heng, M. (1998, March). *Buddhism and the making democracy in Cambodia.* Cambodian Buddhist Council.

Heng, M. (2022). Step by step from Cambodia toward ASEAN. In M. Indrawan, J. B. Luzar, H. Hanna, & T. Mayer (Eds.), *Civic engagement in Asia* (pp. 237–248). Springer. https://doi.org/10.1007/978-981-16-9384-7_22

Hess, C. (2008). Mapping the new commons. *SSRN Electronic Journal*. https://doi.org/10.2139/ssrn.1356835

Jayasuriya, L. (2008). Buddhism, politics, and statecraft. *International Journal of Buddhist Thought and Culture, 11*, 41–74.

Jinananda, B. (1961). *Upasampadājñaptiḥ: Critically edited with notes and introduction.* Kashi Prasad Jayaswal Research Inst.

Keefe, A. A. (1997). Visions of interconnectedness in engaged Buddhism and feminist theology. *Buddhist-Christian Studies, 61–76*. https://doi.org/10.2307/1390399

Keown, D. (1996). Karma, character, and consequentialism. *The Journal of Religious Ethics, 24*(2), 329–350. http://www.jstor.org/stable/40015213

Keown, D. (2016). *The nature of Buddhist ethics.* Springer.

Kostakis, V., & Pazaitis, A. (2020). Who creates value? *Administrative Culture, 21*(1), 76–85. https://doi.org/10.32994/hk.v21i1.253

Lekakis, S. (Ed.). (2020). *Cultural heritage in the realm of the commons: Conversations on the case of Greece.* Ubiquity Press. https://doi.org/10.5334/bcj

Long, W. J. (2019). *Tantric state: A Buddhist approach to democracy and development in Bhutan* (1st ed.). Oxford University Press. https://doi.org/10.1093/oso/9780190843397.001.0001

Long, W. J. (2021). Radical interdependence: Buddhist philosophical foundations for social theory. In W. J. Long (Ed.), *A Buddhist approach to international relations* (pp. 19–33). Springer International Publishing. https://doi.org/10.1007/978-3-030-68042-8_2

Machiavelli, N. (1532). *The prince* (W. Marriott, Trans.). Penguin Classics.

Mandalaki, E., & Fotaki, M. (2020). The bodies of the commons: Towards a relational embodied ethics of the commons. *Journal of Business Ethics, 166*(4), 745–760. https://doi.org/10.1007/s10551-020-04581-7

Mauss, M. ([1925] 1990). *The gift: Forms and functions of exchange in Archaic societies* (I. Cunnison, Trans.). Cohen and West.

Maxwell, T. S. (2007). The stele inscription of Preah Khan, Angkor text with translation and commentary. *Udaya: Journal of Khmer Studies, 8*, 36.

McMahan, D. (2008). A brief history of interdependence. *Pacific World: Journal of the Institute of Buddhist Studies, 10*, 131–176. https://buddhism.lib.ntu.edu.tw/FULLTEXT/JR-BJ011/bj011377398.10.pdf

Ng, E. C., & Ng, E. C. (2020). Interconnectedness, Competition, and Collaboration. In E. C. Ng, *Introduction to Buddhist economics: The relevance of Buddhist values in contemporary economy and society* (pp. 117–130). Palgrave Macmillan.

Ongaro, E., & Tantardini, M. (2023a). Religion, spirituality, faith and public administration: A literature review and outlook. *Public Policy and Administration,* 1–25. https://doi.org/10.1177/09520767221146866

Ongaro, E., & Tantardini, M. (2023b). Debate: Why the religious factor has been forgotten in PA studies? (and how to remedy it). *Public Money & Management*, 1–2. https://doi.org/10.1080/09540962.2023.2264054

Ostrom, E. (2008). Design principles of robust property-rights institutions: What have we learned. *Elinor Ostrom and the Bloomington School of Political Economy: Resource Governance, 2*, 215–248.

Ostrom, E. (2015). *Governing the commons: The evolution of institutions for collective action.* Cambridge University Press.

Pansera, M., & Fressoli, M. (2021). Innovation without growth: Frameworks for understanding technological change in a post-growth era. *Organization, 28*(3), 380–404. https://doi.org/10.1177/1350508420973631

Payutto, P. (1992). *Buddhist economics: A middle way for the market place.* Buddhadhamma Foundation.

Payutto, P. A. (1994). *Dependent origination: The Buddhist law of conditionality* (B. Evans, Trans.). Buddhadhamma Foundation. https://www.watnyanaves.net/uploads/File/books/pdf/dependent_origination_the_buddhist_law_of_conditionality.pdf
Prebish, C. S. (2018). The Vinaya. In D. Cozort & J. M. Shields (Eds.), *The Oxford handbook of Buddhist ethics* (pp. 96–115). Oxford University Press.
Rahula, W. (1959). *What the Buddha taught*. Grove Press.
Reynolds, F. E., & Hallisey, C. (1987). Buddhism: An overview. In M. Eliade (Ed.), *The encyclopedia of religion* (Vol. 2, pp. 334–351).
San, P. (2024). Buddhist governance: Navigating today's role of Saṅgha and Dhammarājā, with special reference to Cambodia. *Halduskultuur, 22*(2), 26–49.
San, P., & Shakya, S. (2025). Buddhist third places: A note on the sattal structure in Nepal and Dharmasālā in Cambodia. *Religions, 16*(4), 478.
San, P., Drechsler, W., & Shakya, S. (2023). Buddhism, wealth, and privilege: Ambedkar and Habermas. *Religions, 14*(8), 1057. https://doi.org/10.3390/rel14081057
Strenski, I. (1983). On generalized exchange and the domestication of the Sangha. *Man, 18*(3), 463–477.
Tambiah, S. J. (1970). *Buddhism and the spirit cults in north-east Thailand*. Cambridge University Press.
Thanissaro, B. (2013a). *Vera Sutta: Animosity.* Retrieved November 2023, from Access to Insight (BCBS Edition): https://www.accesstoinsight.org/tipitaka/an/an10/an10.092.than.html
Thanissaro, B. (2013b). *Maha-parinibbana Sutta: The great discourse on the total unbinding (DN 16).* Retrieved October 2023, from Access to Insight (BCBS Edition): http://www.accesstoinsight.org/tipitaka/dn/dn.16.5-6.than.html
Thanissaro, B. (2013c). *The Buddhist monastic code I and II: The Khandhaka rules translated and explained.* Metta Forest Monastery.
Rhys-Davids, T. W., & Oldenberg, H. (1881). *Vinaya texts: Translation form the Pāli.* The Clarendon Press.
Verma, R. (2017). Gross national happiness: Meaning, measure and degrowth in a living development alternative. *Journal of Political Ecology, 24*(1). https://doi.org/10.2458/v24i1.20885
Weber, M. (1958). *The religion of India: The sociology of Hinduism and Buddhism.* Free Press.
Weber, M. (1988). *The sociology of religion* (2nd ed., Trans. E. Fischoff). Beacon Press.

Pisith San is a researcher at the Ragnar Nurkse Department of Innovation and Governance, Tallinn University of Technology. His interdisciplinary work connects Buddhist ethical and economic perspectives with commons theory and cosmolocalist approaches to governance. He brings a unique perspective that bridges scholarly inquiry with ethical and spiritual traditions in Southeast Asia.

Asimina Kouvara is a researcher at the Ragnar Nurkse Department of Innovation and Governance, Tallinn University of Technology. Her work lies at the intersection of design, open fabrication, and sustainable production, exploring how distributed manufacturing and co-creative processes can foster convivial, locally rooted yet globally connected built environments. She contributes to research on commons-oriented and cosmolocal approaches to technology and community-driven innovation.

Open Access This chapter is licensed under the terms of the Creative Commons Attribution-NonCommercial-NoDerivatives 4.0 International License (http://creativecommons.org/licenses/by-nc-nd/4.0/), which permits any noncommercial use, sharing, distribution and reproduction in any medium or format, as long as you give appropriate credit to the original author(s) and the source, provide a link to the Creative Commons license and indicate if you modified the licensed material. You do not have permission under this license to share adapted material derived from this chapter or parts of it.

The images or other third party material in this chapter are included in the chapter's Creative Commons license, unless indicated otherwise in a credit line to the material. If material is not included in the chapter's Creative Commons license and your intended use is not permitted by statutory regulation or exceeds the permitted use, you will need to obtain permission directly from the copyright holder.

Knowledge Ethics for Marine Spatial Planning: Balancing Critical and Constructive Engagement with Small-Scale Fisheries

Svein Jentoft

Abstract Through territorial demarcation and allocation, Marine Spatial Planning (MSP) aims to bring order and direction into an increasingly crowded, chaotic, uneven, and contested marine commons. To achieve sustainable and equitable outcomes, I argue that MSP must explore how the commons form a socio-ecological and political system. This can only be achieved through interdisciplinary and transdisciplinary governance approaches where people with local, practical, and ethical knowledge play a part. This chapter is concerned with the fate of small-scale fisheries that are vulnerable in the Blue Economy and what MSP can do to secure their livelihoods and communities. When MSP is a process of transdisciplinary co-production in which small-scale fisheries people play a part, they do not only provide access to what they know about the commons as a system. Planners also get to know about their values, ethics, worldviews which they have acquired a from being on the water and from growing up in a small-scale fisheries community. MSP will not succeed unless the latter is considered relevant in the planning process.

> States should, as appropriate, develop and use spatial planning approaches, including inland and marine spatial planning, which take due account of the small-scale fisheries interests and role in integrated coastal zone management.
>
> Voluntary Guidelines for the Support of Sustainable Small-Scale Fisheries.
>
> (SSF Guidelines, paragraph 10,2, FAO 2015)

S. Jentoft (✉)
Norwegian College of Fishery Science, UiT-The Arctic University of Norway, Tromsø, Norway
e-mail: svein.jentoft@uit.no

© The Author(s) 2026

S. Partelow (ed.), *Ethics and the Commons*, Schriften zu Leben und Gesellschaft - Series on Life and Society 3, https://doi.org/10.1007/978-3-032-24246-4_6

1 Introduction

Small-scale fisheries are small in scale but large in numbers. Globally, they employ and feed hundreds of millions of people, but exist in small, localized configurations with high social and ecological diversity (FAO et al., 2023). For academics, small-scale fisheries provide a case for reflecting on the role scientific and practical user knowledge plays in shaping social and ecological realities towards sustainable livelihoods and social justice. This includes exploring the way knowledge is supposed to play - and could possibly play - in marine governance and marine spatial planning (MSP) in the Blue Economy.[1] Lessons from research on small-scale fisheries have implications for knowledge production practices beyond them, also in other sectors where science and societal actors are involved in the production of knowledge for sustainable governance.

As I will argue in this chapter, small-scale fisheries people have a substantial amount of what Aristotle refers to as *phronesis* or practical wisdom, which needs to play a part in coastal governance (Jentoft, 2006; Caniglia et al., 2023). Enabling small-scale fisheries and fisheries communities to participate in MSP is both a moral imperative and an institutional measure that can improve how such processes can move beyond technical implementation towards balancing the role that diverse knowledges and values can play in shaping sustainable and equitable outcomes. **MSP** is defined by UNESCO as "a public process of analyzing and allocating the spatial and temporal distribution of human activities in marine areas to achieve ecological, and social objectives that are typically specified through the political process."[2] How MSP works or can possibly work as a process of co-production in concrete settings is an important question, especially for affected small-scale fisheries for whom the outcome may come with both benefits and costs.

In what follows, the contribution of science in MSP is discussed in sect. 2. The argument here is that there is a need for balancing critical and constructive perspectives. For small-scale fisheries, the Blue Economy comes with new opportunities but also risks. Section 3 explores how scientists may help to clarify whether MSP can harness the risks that the Blue Economy pose for traditional and established users of the marine commons, most prominently small-scale fisheries people, and their communities. Section 4 reflects on the way scientific language plays in framing these opportunities and risks. The focus is on how metaphors shape the perception of challenges and determine goals in MSP. Metaphors, as Plato noted (Jentoft, 2023a), are a weaker form of knowledge than what is achievable through empirical research and the experience-based knowledge of the coastal commoners in MSP, like that of small-scale fisheries people. Section 4 examines transdisciplinarity as knowledge co-production and how the ethics and morality of small-scale fisheries people must be a knowledge dimension in MSP.

---

[1] Blue Economy: oceans as the next great economic frontier (unric.org)

[2] https://www.ioc.unesco.org/en/marine-spatial-planningeconomic

2 The Role of Science

For scientists interested in MSP and small-scale fisheries, there is a need to balance critical and constructive discourse and to move beyond disciplinary approaches to foster pluralism and collaboration in the form of inter- and transdisciplinarity. Scientists must reflect on their contribution, including how their language metaphors frame the understanding of what is at stake. Language is not just a tool for representing realities in the Blue Economy, but also for shaping it.

Scientists are making a positive difference in the transformation that must occur to support and sustain the lifeworld and wellbeing of small-scale fisheries people. This they must do in the face of major threats, be they climate change, resource degradation, poverty, and hunger and so forth, as specified in the seventeen UN Sustainable Development Goals.[3] Then, scientists need to be conscious about how to balance goals from a critical and constructive perspective. In the case of MSP in the Blue Economy, they should not just point at the potentials to secure the survival of small-scale fisheries but also at the risks of further marginalizing small-scale fisheries and their communities. Small-scale fisheries have many things at stake in the Blue Economy, including their freedoms and human rights (Willmann et al., 2017).

However, that may involve a role conflict. Being critical and constructive may be difficult at the same time. Still, to be constructive, scientists must also be critical. To get it right, they must also discuss the chances of getting it wrong. In his book, "Making social science matter", Flyvbjerg (2001), argues that social scientists should always ask where society is heading and whether this is desirable. There are also issues of equity and justice that must be addressed: Who gains and who loses, by which mechanisms of power? What should be done when on the wrong track? These are questions that should also be examined in the context of MSP. They are also relevant in the context of the wider Blue Economy, in which MSP is supposed to be a transdisciplinary tool for providing an integrated, holistic, and balanced assessment of marine and coastal commons value- and interest-conflicts.

Critical and constructive perspectives are especially important when discussing schemes that can lead to government policy reform. For example, despite the best of intentions, the government may not have considered all the externalities and risks. Their approach may have been different if they had. In this case, the job of scientists would not just be to help government do a better job with MSP but to help small-scale fisheries cope in a situation where they are vulnerable. MSP is not "a walk in the park", as the saying goes, but a challenging undertaking as the growing literature on its merits demonstrates (cf. for example Jones et al., 2016; Trouillet, 2019; Flannery, 2023). What MSP can do for small-scale fisheries in the Blue Economy begs for both critical and constructive analysis, as I will try to explain here.

MSP would also benefit from the reflective analysis of the limitations of science, how science can fill the knowledge gaps and thus be a measure against governance failure. Planners may have to look for other sources of knowledge than what science can produce. This raises the question of what kind of knowledge is needed, and what

[3] THE 17 GOALS | Sustainable Development (un.org)

different knowledges may contribute to producing better outcomes in the marine realm, also for small-scale fisheries people. Notably, the limits of knowledge in producing positive outcomes for small-scale fisheries are not just within the sciences but also external to them, as within the structures of power in the Blue Economy (Garland et al., 2019; Schutter et al., 2021).

At the end of the day, the fate of small-scale fisheries would be determined outside academia, regardless of the quality of scientific contributions. MSP may be victim of power grabbing from stakeholders who see it in their interests to limit what it can possibly do for small-scale fisheries and enhance what it can do for themselves. This, however, should not discourage scientists of different disciplines to engage, preferably through a joint effort, to clarify what the stakes are. This chapter posits that to make the best job possible, scientists must also welcome the knowledge contributions of small-scale fisheries people, including their ethics and morality. Therefore, they must aim beyond interdisciplinarity, for transdisciplinarity (Chuenpagdee & Jentoft, 2018). This is especially important with the marine and coastal commons as a shared space of multiple, often conflictive stakeholder interests, ambitions, and worldviews.

3 The Blue Economy

3.1 Blue Justice

Small-scale fisheries are facing increasing competition for access to space and resources in the coastal zone. They cannot any longer take for granted that they have the same space and resources available to them that they relied on, often for generations. The Voluntary Guidelines for Securing Sustainable Small-Scale Fisheries (SSF Guidelines), which FAO (UN Organization for Food and Agriculture) member states endorsed in 2014, recognize this growing threat.

> "Small-scale fishing communities also commonly suffer from unequal power relations. In many places, conflicts with large-scale fishing operations are an issue, and there is increasingly high interdependence or competition between small-scale fisheries and other sectors. These other sectors can often have stronger political or economic influence, and they include: tourism, aquaculture, agriculture, energy, mining, industry, and infrastructure developments." (p. x)

This observation finds support in the research literature, like that of Janßen et al. (2018), Trouillet (2019), Tafon (2019), Flannery et al. (2019) and Das (2023). It is also the topic of a recent book I edited together with colleagues which introduced the concept of "blue justice" in the context of the Blue Economy (Jentoft et al., 2022), where the problem that the SSF Guidelines describes is likely to increase because of the entrants of new actors entering the coastal zone in search of new business opportunities, in what is often described as the new frontier. As the less powerful stakeholder, small-scale fisheries are facing the risk of being overrun, pushed aside, further marginalized, and even eradicated. The Blue Economy thus raises serious ethics and justice issues. They are not new, but are likely to become more prevalent with Blue Growth policies.

The Blue Economy and Blue Growth policies involve "positive goals", i.e., things policymakers aspire to accomplish, and which MSP, if done right, may help achieve, such as food security, secure livelihoods, thriving local communities, and gender equality, to mention a few of them. "Negative goals" refer to things to avoid, like overfishing, environmental degradation, further marginalization of small-scale fisheries and their communities, that may lead to the erosion to customary rights, cultural heritage or social equity. To be the tool that policy makers hope it will be, MSP must aim to achieve both goal types through a process that is democratic and inclusive of small-scale fisheries people (Jones et al., 2016). The Blue Justice concept encompasses both positive and negative goals, like those mentioned here.

As Flyvbjerg posits, it is naïve to assume that economic and social change only create winners and no losers. This is also a caveat for the Blue Economy. The win-win for all stakeholders is not realistic. Researchers and planners alike must critically assess what takes place in concrete settings, who the winners and losers are. It would contradict our sense of social justice and indeed important governance principles if the poorest and most powerless among marine stakeholders are those who must carry the heaviest burden in the Blue Growth transformation. MSP research should therefore investigate how small-scale fisheries people end up, and whether MSP is inclusive or exclusive of their interests and concerns (Flannery et al., 2019).

3.2 Blue Conflict

When the Blue Economy brings new stakeholders into the coastal commons, conditions for exhausting the carrying capacity of natural resources and for generating the potential of more social conflict is likely to increase. This will not only threaten livelihoods of local people, but also hamper growth. Who would invest in an area that looks like a "warzone"? More than anything else, investors want stable and hence predictable working conditions. MSP is meant to establish such conditions and by that encourage economic growth (Eskeland Schütz & Slater, 2019). Whether MSP can deliver this, is an empirical question. Does it ease or exacerbate pressure on marine resources, does it put conflicts at rest or add fuel to them? With geographical organization and spatial demarcation, does MSP fence small-scale fisheries operators in or out? Neither is good for small-scale fishers who need autonomy to be flexible. Is allocation of space fair to small-scale fisheries, their tenure rights, and their perception that the coastal commons is as much place as space?

In my own country, Norway, stakeholder conflicts are daily news. There are large-scale against small-scale fisheries, small-scale fishers against tourist fishers, against fish farming, fish farming and small-scale fisheries against marine conservation, like marine protected areas (MPAs), marine windmill parks against fisheries interests, oil exploration vs. fisheries and environmental concerns, toxic wastewater from mining running into the fjords that pollutes the marine ecosystem, and so forth. Norway has a relatively good planning system at all scales, but it does not effectively secure small-scale fisheries from the pressures of powerful interests. They are involved in an ongoing struggle to defend their position, in the media and on fishing grounds. In 2024, a reform-proposition of the quota system triggered heavy opposition from small-scale fishers by shutting off harbors so that no one could come in our out.

Researchers must examine how small-scale fishers respond to the challenges and opportunities that the Blue Economy brings. Are they excluded from, or actively participating in MSP? They must also critically examine the rhetoric. Blue Growth language may promise more than it can deliver. The Blue Economy *is* what it does, not what it says. Concepts get their true meaning in the real world, in their use, as Wittgenstein (1953) argued, and not at someone's desk or in political propaganda. Should MSP erode the commons by excluding small-scale fishers and turn coastal space into private property of corporate stakeholders, it would tell us what MSP is and does, even if it is not among its primary goals but a latent function. One cannot say that further marginalizing small-scale fisheries is not what MSP is, if that is what happens because of it. One cannot expect to convince small-scale fisheries people by wrapping failures in euphemisms.

4 Metaphors to Govern by

4.1 Images

The knowledge about the marine social and ecological system is often shallow and patchy. Therefore, planning authorities are shooting in the dark. They have limited understanding of what they are stepping into. In the absence of empirical knowledge, they are often backing their policies with metaphors, like Garrett Hardin's the Tragedy of the Commons (Hardin, 1968). The essence of the metaphor is to "understand and experiencing one thing in terms of another" (Lakoff & Johnson, 1980, p. 5). Here lies the risk of misrepresenting the reality that the metaphor is meant to capture.

Hardin asks the reader of his paper to "picture" a pasture with freely grazing cattle and with no limit on the size of the herd. Simple logic predicts that overgrazing will happen with a dismal result. When the freedom of the commons remains, he argued that a situation akin to the ancient Greek theatrical plot is inevitable. Hardin's metaphor is frequently applied to justify government intervention into the marine realm. It conveys a message of caution and urgency; the need to act before it is too late. When the collapse of the commons is a fact. With this pressure on policy makers and managers, research is a waste of time. Forget about the collateral damages. One may have to break some eggs; the omelet is a common good. But to stay with the metaphor; if you are among the eggs, like small-scale fisheries people often are in MSP, you will think differently. In a zero-sum society, as Thurow (2001) pointed out, there is little comfort to the loser that the collective gain exceeds his loss. With

MSP, small-scale fishers have reason to ask what MSP is doing for them and their community.

When in the Blue Economy, the state intervenes into the marine commons to escape the Tragedy of the Commons, it threads into a complex social and ecological system that already works much as a zero-sum society. Hardin's metaphor conceals this reality. The pasture that he describes looks like a marketplace, not a community, but an "interdependent" rather than a "functional system", as Boudon (1981) coined them. The commoners in Hardin's allegory invest for private rather than collective gain. They have no concern for each other as commoners, or for the quality of the pasture as a commons. They are only in it for themselves.

Marine planners should not move blindly into the new Blue Economy era. From a justice perspective, MSP should guarantee that small-scale fishers, as the weakest stakeholder group, do not lose their fishing grounds, beaches, and homesteads. The commons have traditionally served small-scale fisheries and their local communities well, but now they need legal protection against encroachment by newcomers. According to FAO (2015), their traditional tenure systems must be respected as a human right. Whether MSP will deliver on this necessity, is a crucial issue for their sustainability in the Blue Economy. If not, they have reason to fear it.

4.2 A Level Playing Field

The Blue Economy language is also metaphorical, as when the European Union says that it will continue to support a "level-playing field" in the Blue Economy.[4] Likewise, the metaphor appears in MSP research papers (Keijser et al., 2018: Smith & Jentoft, 2017). Then, the idea seems to be that marine stakeholders are involved in a friendly sport, one about winning and losing, like in football where the ethos of "fair play" reigns. A tilted coastal playing field would be unfair to those that have to play "uphill". Therefore, in the name of Blue Justice, the task of MSP would be to level it. In football, the field is balanced and even, rules are known and apply equally to everyone. Leonel Messi plays within given, well-established rules, which apply to him in the same way as they do to all players on the pitch. Knowing football requires knowing the rules but also the play—what is going on.

This, however, is not how the coastal commons appears in the Blue Economy, where rules are not given and stabilized, but part of the play. Powerful players develop their own rules to serve their interest. They may succeed in getting planning authorities on their side. Then, Blue injustice is predictable, especially for small-scale fisheries playing in a lower league. Their players that do not possess the resources and capabilities to fix the rules and win the game. Football is not a metaphor that sufficiently represent the structure of, and interactions in, the marine commons in

[4] Creating a level-playing field based on decent work and social sustainability in the blue economy | Department of Economic and Social Affairs (un.org). See also 2022-blue-economy-report_en.pdf (europa.eu)

the Blue Economy—which, by the way, is itself a metaphor. The economy has no color.

Whether MSP can withstand corporate power and secure the rights and interests of small-scale fisheries is a question for empirical investigation. If MSP fails to do so, the marginal position and dismal fate of small-scale fisheries is sealed. Unequal power-relations work against leveling the coastal playing field. Small-scale fishers do not have "coaches", like lawyers or political consultants, to represent them. They often lack basic organizational structures and capacities and are therefore not playing as a team. Instead, they are left to fend for themselves. They may not even know what game they are in, what their stakes are, and what it takes to win. No one is there to tell them. Here, social scientists have a constructive role to play, but primarily after they have critically examined how the game works and who is involved.

The level playing field metaphor is disarming to small-scale fisheries. Hardin's image of the "tragedy of the commons" works the same way. In the Blue Economy, metaphors should urge critical investigation of social justice, the function of language, and whose ethics prevail. Metaphors must inspire nuanced, critical analysis, also of discursive power. Here, Pierre Bourdieu has metaphors to offer.

4.3 Habitus and the Social Construction of Commoners

Like Hardin, Bourdieu (2005) found[5] inspiration in ancient concepts. His "habitus" concept, for instance, is from Aristotle. In simple terms, it refers to values, characters, abilities, and habits that a person or a group have acquired though experience and have bearing on how they orient themselves in the world. Bourdieu argues like Marx before him, that people in their agency simultaneously reflect and reproduce social structures and therefore also themselves. He says that "*habitus* defines the perception of the situation that determines it" (Bourdieu, 1995, p. 87).

As mentioned, in the context of MSP, the stakes of small-scale fishers should be examined. But one should also clarify who the other stakeholders are and how they become who they are, in other words their habitus. One should examine how their habitus works in planning settings, and how the planning process is changing it. Hardin assumes a rational agent, a homo economicus, who acts in self-interest. But, following Bourdieu, whatever the agent was to begin with, the "play" forms him or her. The intervention to prevent the tragedy or level the playing field creates a new situation that transforms the game and in the next instance the actors. If actors were not homo economicus at the outset, they may now become like him. Here is Bourdieu (2005) again:

> "The most basic economic dispositions - needs, preferences, propensities – are not exogenous, that is to say, dependent on a universal human nature, but *endogenous* and dependent on a history that is the very history of the economic cosmos in which these dispositions are required and rewarded." (p. 8).

[5] Parts of this and the following section are drawn from my book (Jentoft 2023)

Marcel Mauss (1954), another French sociologist, puts the same observation into simpler language:

> "*Homo economicus* is not behind us, but before, like the moral man, the man of duty, the scientific man and the reasonable man. For a long time, man was something quite different; and it is not so long now since he became a machine - a calculating machine." (p. 74).

In other words, homo economicus, as Hardin describes the commoner, is not a natural but a social construct. Hardin uses a metaphor that reduces his habitus into a narrowly focused, economically rational individual. Bourdieu's metaphor alludes to a richer, and more complex actor. Regardless of how the commoner are to begin with, the question that should be asked is therefore if the habitus of small-scale fishers and other stakeholders must become like "calculating machines" to survive in the Blue Economy, as if that is the only relevant characteristic.

In small-scale fisheries, the argument is already out (Maurstad, 2000). The quota systems that are now being introduced throughout the European Union and around the world are turning fishers into "calculating machines." The sub-title of Jeppe Høst's book (2015) on the Danish quota-system "Captains of Finance" is telling. It is a metaphor originating from the Norwegian/American economist Torstein Veblen, which in Høst's narrative indicates a fundamental transformation of the fisher habitus.

Fishers used to regard themselves and by others as hunters, free but raised and socialized within local communities, not the culture of financial markets. That may now be changing. How close or distant the habitus of coastal commoners, like small-scale fishers, is from the *calculating machine*, may be subject to empirical investigation. In the Blue Economy they are encountering other stakeholders who were raised differently and for whom their presence in the coastal commons is contrarily motivated. Social scientists should explore how the habitus of small-scale fisheries people and other coastal actors is both affecting and is affected by MSP.

But social scientists should be skeptical of their own assumptions, as the Greek philosopher Sextus Empiricus advised, and whose name gave inspiration to scientific research. Computer games provide insights in the functioning of the marine commons, and they are played in a literal sense.[6] The Prisoner's Dilemma, which captures Hardin's logic, is the same. Metaphors, on the other hand, are constructed in the mind, whereas reality must be experienced. Contrary to what scientists and MSP assume, small-scale fishers may still have the old ethos and ethics intact. The Blue Economy may not yet have totally transformed their habitus but may well be in the process of doing so.

In MSP, sociologists must look for 'self-fulfilling prophecies', as Robert Merton (1948), and Bronislaw Malinowski, (1922) before him, coined them. These are consequences no one planned for, as when assumptions unintendedly become reality when acted on. Then, metaphors are validated in a process, resulting false learning. MSP would turn small-scale fishers into calculating machines you believed were natural in the first place, with the Tragedy of the Commons as a given outcome. Herein lies the danger of metaphors, as Ellinor Ostrom (1990) pointed out. When acting on them,

[6] https://rmitchel.uoregon.edu/commons

they become social facts, and MSP authorities get what they need to justify their intervention.

4.4 Social Field

Bourdieu borrowed the concept of "social field" from physics (Hilgers & Mangez, 2015), when talking about society as divided into multiple social fields, like the economy, the polity, the arts, and sports, etc., and which are autonomous domains with actors of different *habiti*. People may be involved in several fields, but the habitus one may possess in one field is not relevant to another. For instance, you do not become a good fisher as a good footballer, or vice versa. From a Bourdieu perspective, the Blue Economy would be perceived as an emerging new social field—or an assemblage of multiple sub-fields, with small-scale fisheries being one of them.

For Bourdieu, social reality is relational, and a field like the commons is a social system. He assumes that relations shape interaction, and in the next instance the habitus, in an iterative process. By following his line of thought in the context of MSP in the Blue Economy, a complex image of the coastal commons appears. When searching the internet for images of Bourdieu's concept of social field, one gets something very different from the looks of a field where footballers play or where grazing cattle would appear.

MSP would be much easier if the social field on the coast were akin to a pasture or a pitch. Bourdieu's concept suggests a more complex task. It entails a social process that challenges the capacities and capabilities of planners and stretches the limits of MSP as a governance instrument.

5 Crossing Disciplinary Boundaries

5.1 Semantic Infiltration

A deeper understanding of the structure and functioning of the commons as a social field in the Blue Economy, would result if different academic disciplines were working better together. Multi-disciplinarity would be a step in the right direction, but interdisciplinary would take us further. However, interdisciplinarity is easier said than done. Scientists do not communicate well across disciplinary boundaries. Disciplines are not called disciplines for nothing. They are institutions with strong norms for how and why they do science (Buanes & Jentoft, 2009). Bridging established paradigms is not cost-free for those who do it. Their heterodoxy may have negative consequences for their career.

There are also issues concerning basic paradigms and assumptions about how the world works and what drives people deep down. Are they natural "calculating machines" or are they complex social beings striving to make sense of themselves and the situation they are in. Then there is the language issue. Disciplines are not just defined by their topics and methodologies, but also by their conceptualizations. For instance, Bourdieu is among the most cited social scientists in the world, but not well known outside sociology and anthropology. Mention the habitus concept to an economist or a biologist, and they would not understand what you are talking about. Their own disciplinary vocabulary will meet similar confusion with sociologists or anthropologists.

However, although disciplines have special areas of interests, they also have overlapping concerns, which should make them interested in talking together. The functioning of the resource commons draws widespread interest. But they would have some translation to do to fully grasp each other, which does not only require pedagogical skills but also skills in listening. Interdisciplinarity may easily evolve into a contest in whose knowledge is superior (Sennet 2012). The sociologist and US senator Patrick Moynahan invoked the concept of "semantic infiltration"[7] It refers to a situation where you force one person to use your language rather than his own. Such a behavior would distort interdisciplinary and transdisciplinary communication, making learning a one-way rather than an interactive process. (Visser, 2004).

Nonetheless, even if scientists should succeed in achieving interdisciplinarity, there is still a way to go before they get a fully holistic idea of how the commons work. The sciences have more to offer MSP if they work together than when isolated from each other. However, science is a special way of knowing and does not have all it takes to know everything. Transdisciplinary approaches look beyond them (Chuenpagdee & Jentoft, 2018).

Lang et al. (2012, p. 26) define transdisciplinarity as "a reflexive, integrative method driven scientific principle aiming at the solution or transition of societal problems and concurrently of related scientific problems by differentiating and integrating knowledge from various scientific and societal bodies of knowledge." Transdisciplinarity integrates the sciences but also draws on the knowledge that non-scientists have, like that of small-scale fishers. Fishers make observations of what is happening around them, also below the water. They make inferences from their experiences, about the "differences that make a difference", as Gregory Bateson (2000) phrased it when he talked about local knowledge.

5.2 *Localized Knowledge*

It seems generally accepted now, also among natural scientists, that local ecological knowledge is relevant to science, management and planning in the marine commons. However, the presumption is not that local knowledge is necessarily reliable. Fishers may not always see what they believe they are seeing (Davis et al., 2004). Neither do scientists. False learning occurs in both camps. Non-scientists, like small-scale

[7] Un-words and Policy - The New York Times (nytimes.com)

fishers, may well be right when scientists are wrong. Besides, members of the same camp do not always agree on what they are observing, and they may have different ideas of what the causes and consequences are. In both instances, reality tests would be in order. In Norway, fishers long argued that the Arctic cod, that visits our shores every winter to spawn, is a different species than the cod that resides and spawn in the fjords. They claimed that with their trained eye they could see the difference between the cod species. Now, marine scientists have confirmed that fishers are right, and that the fjord spawning grounds consequently need special protection, for instance from aquaculture installations. For MSP, this is a major revelation.

Local ecological knowledge used to be rejected as anecdotal knowledge, just stories, which could not be trusted. A rather arrogant view, of course, which now seems to have died out. Today you sound foolish if you draw the anecdote card. Bob Johannes's book "Words of the Lagoon", which came out in 1981 based on his work on the Palau islands in Micronesia, helped to change this attitude. Just as there is a folk medicine, there is a folk ecology and biology, and it should not be rejected out of hand. Neither should their folk sociology and, as I will argue here, their folk philosophy and their ethics.

Local ecological knowledge has become a popular research issue. Numerous studies (cf. for example (Dyer & McGoodwin, 1994; Berkes, 1999; Haggan et al., 2007) for early contributions) show that experienced small-scale fishers have extensive observational knowledge of complex system dynamics, often passed on over generations in the form of stories and ritual. Like all observations, they are prone to bias and subjective interpretation, but they are what small-scale fishers act on. Nevertheless, they may be onto something that should be systematically explored by science. Their experience-based knowledge are at least hypotheses for scientists to evaluate.

5.3 Ethics

As an ethical principle, MSP should respect small-scale fishers for what they know about their marine environment and resources from being on the water (FAO, 2015). This is also a matter of research ethics that requires valuing detailed, unbiased observations of complex systems that has become embedded into the operational and cultural lives of fishers. Without this ethics, transdisciplinarity is deemed to fail.

A researcher who wants to know what local people know, cannot just go and pick it up, like a botanist would do with flowers in the field. Local knowledge is not easily available, for instance as written text. Neither is it always contained in the stories people tell. Stories are sometimes told to mislead to protect the resource, or only contain fragments of fact. People have all kinds of beliefs in things they do not really understand or want to share. Sometimes they hold them to be sacred, like Sunde (2014) describes from a South African fisheries community where people go to rocks in the sea to speak to ancestors—until a new Marine Protected Area fenced them out. In his book Sacred Knowledge, Fikret Berkes (1999) writes: "Traditional knowledge systems tend to have a large moral and ethical context, there is no separation between nature and culture. In many traditional cultures, nature is imbued with sacredness (p. 12).

Local knowledge may not be for the outsider. MSP may require that local knowledge be exposed which fishers prefer to keep for themselves. Local knowledge is also tacit, embedded in practice, in the things people do, in what they pay attention to, in their skills. There is local knowledge in the gear fishers use and in the design of their boats. A professional fisher does not need to consult a manual when fishing. The knowledge is already in his body, in the way he moves, like how he uses his knife and maneuvers at sea. Like a scientist, he reads nature like a book. He needs to be his own biologist, meteorologist, and engineer, and he must also know how to run his business, steer his boat, and lead his crew.

Local knowledge is something that small-scale fishers have about the local situation and the particular condition under which they work. Thus, it is localized knowledge, not necessarily relevant or sufficiently accurate and relevant for other locations. Norwegian fishers may know their fjord well, but not the fjord next to theirs if they do not fish there. Their local knowledge is not easily transportable to another locality; they cannot just go anywhere and apply their knowledge in a new place. MSP is therefore not only about space, as the term suggests, but also place. MSP is not about drawing lines on a map, but intervention into an area of cognitive significance to local users.

A researcher may get access to this knowledge by engaging with the fisher, by being there, joining him when he goes fishing, by watching what he does, talking to him, learning his idioms. Thus, a fishing boat can be a research vessel. The fisher will show you things, tell you stuff, even confide in you if you have his trust. He will also explain what is not intuitively clear about what he does and why he does it. He is the expert; you are the novice. You learn and listen carefully and respectfully, and record what he says and does. Some things you would need to figure out yourself. Then you have data to bring home for further analysis. This is what it means to do anthropological fieldwork in small-scale fisheries.

The question I shall discuss next is whether local knowledge is just about a different set of data, about empirical facts that scientists and managers can collect. Or is it more to local knowledge and transdisciplinarity that is important to MSP?

5.4 *Morality and Law*

Garrett Hardin is criticized for misrepresenting reality with his metaphor about the Tragedy of the Commons. The habitus of the commoners is fuller than he envisages, and the social field they operate within are more composite and complex than the pasture he talks about. Commoners are not isolated individuals free from social obligations. However, Hardin said one thing that has escaped criticism.

> "And implicit and almost universal assumption of discussing in professional and semipopular scientific journals is that the problem under discussion has a technical solution. A technical solution may be defined as one that requires a change only in the techniques of natural sciences, demanding little or nothing in the way of change in human values or ideas of morality (Hardin, 1968, p. 1343).

The issue worth checking is whether, in the Blue Economy, MSP is working from the same assumption that Garrett Hardin is criticizing here. Is MSP perceived as a technical instrument, only requiring the knowledge of natural science? Or is it also challenging "human values and ideas of morality', something that the social sciences and the humanities would be interested in?

Demarcation and allocation of marine space require mapping that a Geographical Information (GIS) expert can help with. Where to draw the lines on the map needs intimate knowledge of the socio/ecological field, who operates there, and how they connect and interact. This is needed knowledge when looking at the coastal commons with the lens of Bourdieu. Notably, also Bourdieu's field metaphor is a 'map' and not the territory. Gregory Bateson (2000) warned about the mixing of the two. That, however, is precisely the criticism that has been raised against Hardin's Tragedy of the Commons. As an analytical device, the metaphor is just the camera, not the object being photographed, as Erwin Goffman noted (1986).

When recognizing that MSP and GIS are not just technical instruments, but also involves issues that are ethical and moral, how should MSP/GIS account for them? Scientists, and the planners who consult them, have no special authority on values and morality. They are, however, expected to know the law, but law and morality are not separate spheres.

Law is about entitlement, like tenure rights of small-scale fishers, as well as rectitude. It is founded on ideas about right and wrong. Small-scale fishers may have legal rights of access to the resource commons, but there is also a moral reason why they have them. Law and morality are not fully overlapping. There are moral norms that are not legally codified. You can do things that are morally condemnable but not illegal, like being rude to other people, like those you encounter on the fishing ground or in the community. In a Venn-diagram, the relationship between morality and law is portrayed in Fig. 1.

All laws as have an implicit moral position, but not all moral norms are formally codified in law. Moral norms, as illustrated in Figure 1, also exist outside the inner law circle. In law-school, you hear more about the inner than the outer circle. Moral norms

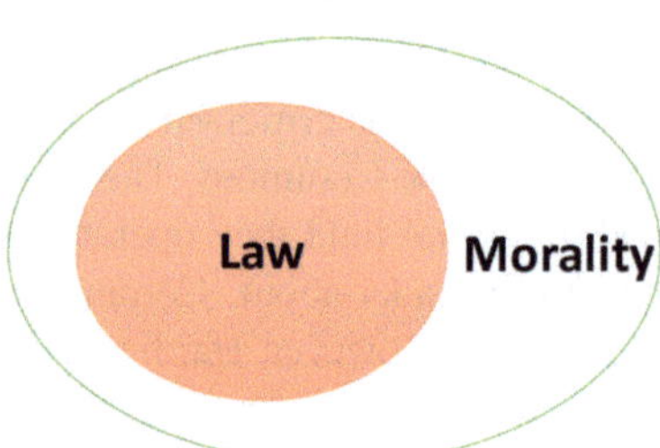

Fig. 1 Law vs. morality (*Source* Jentoft, 2023b). The image depicts a simple circle (law) inside a larger circle (morality)

are something you acquire at home in your family, by growing up in a community, when fishing with your father. There you learn how to behave, how to be grateful and respectful, how to care and be trustworthy.

The social anthropologist, Stephen Gudeman (2001) argues that distinguishing between the commons and the community is a fallacy. The commons should be thought of as a symbol of the community. "The so-called "tragedy of the commons", which refers to the construction of a commonly held resource through unlimited use by individuals, is a tragedy not of a physical common but a human community." (p. 28).

It is the job of MS planners and bureaucrats keep track of the law, to know when legal boundaries are transgressed and what legal entitlements exist. But how do they get access to the morality outside the inner circle, including that of the community?

5.5 *Ethical and Moral Knowledge*

To make MSP appreciative of the ethical and moral aspects of planning, planners would benefit from the input from of the social sciences and the humanities. However, unlike the natural sciences, these sciences do not have much to offer in terms of universal insights, but have a trained eye for contextual factors, like the cultural values and ethics that are underpinning the social interaction among small-scale fisheries people in their local communities. Therefore, Flyvbjerg (2001) posits that the natural, social sciences, and the humanities may build on each other's strengths and compensate for each other's weaknesses. He argues that natural sciences are strong where social sciences are weak, and vice versa. Together they may achieve things that they cannot on their own. With transdisciplinarity, they may go even further and deeper.

Like scientists, small-scale fisheries people learn things about the marine ecological and social field from inhabiting and working in it. They also draw on what people before them have learned. Local knowledge is contained in local communities. It is part of their cultural heritage. Small-scale fisheries people get access to local knowledge by being part of the community. They would not be able to operate without this knowledge. They need to know what they are doing where they are. To survive in an environment that also provides existential threats, they simply must know what the opportunities and risks are. Small-scale fishing is among the most dangerous occupations there is.

But small-scale fisheries people also have values, norms, and principles on how to live and how to make their communities change for the better. These are yardsticks for their judgement about the merits of MSP. Given MSPs intervention into their livelihood opportunities, their voice should be heard as a matter of principle, not just because they have relevant data to bring to the decision-making table. They have a moral right to be proactively part of the planning process. MSP should therefore work as a platform for them to convey their local knowledge but also to express their normativity. The principle of "free, prior, and informed consent" as enshrined in the

UN Declaration of the Rights of Indigenous Peoples", is an ethical principle of wide implication, also in the context of small-scale fisheries (FAO, 2015).

People have moral principles, and they have ideas about how they should be contextualized and implemented in their lifeworld. They do not only have factual knowledge about the system they are part of; they also have wisdom, and they have their own metaphors to express it with. People are experts in their own life. They do not need to consult Aristotle about virtue ethics or Immanuel Kant on the Categorical Imperative about how to relate to others. They know what is morally right and how to behave accordingly. They know what they owe other people and how to meet strangers because that is how they have been brought up.

Small-scale fisheries people are of course no different. Like other people, they learn from early on how to act morally. By growing up in a fisheries community, they have ethical and moral knowledge before they start fishing on how to interact with colleagues at sea. They may compete for the biggest catch, but competition occurs within known moral limits. For instance, you do not put your gear on top of other fishers' gear, or cut the gear of comrades. They communicate and exchange information and help each other when needed. To be safe, they must cooperate. Local fleets are not just made up of vessels that randomly aggregate on the fishing grounds. They form social groups, typically with fellow community members and fisher colleagues they know and have learned to trust and respect (Cf. for instance Acheson, 1998). Norwegian small-scale fishers have the expression of "comrade boats", which are a term for boats traveling and fishing together.

In a transdisciplinary MSP approach, local knowledge - the things small-scale fisheries people know and value - is not only a data supplement to science. Planners also need the knowledge type that Aristotle called *phronesis* or practical wisdom, which is what they get when involving local people and their communities in the MSP process in a co-management arrangement (Jentoft, 2006). Then, planners achieve access to what they know about the marine social and ecological system they are part of and have experience from - if they are willing to share it. They also in a direct way, i.e., "from the horse's mouth", get insights into their wisdom, their values, and moral principles, and their ideas of what constitute social justice in the Blue Economy.

Phronesis is one of the three knowledge forms that Aristotle names in his Nicomachean Ethics. Following Hardin, MSP should not be thought of as a technical, "post-political" process (Flannery et al., 2018), one that is just about the first two knowledge types of Aristotle, episteme and techne. Without his third knowledge form, phronesis, MSP is set up for failure. However, to acquire it requires a different process than that of science (which is what he calls *episteme*), one of participation, liked what Patsy Healey, inspired by Habermas (1984) termed "communicative planning" (Healy, 1996). One must then be prepared for a less linear planning process than how textbook flowcharts depict it. This, however, is how participatory democracy works.

Without mutual understanding and trust, such a process will not work. It requires a readiness for listening, also to the subtext (Sennett, 2012), of what local small-scale fisheries are saying in a language that is not always easy to comprehend for an outsider. Transdisciplinary communication and learning are a complex process involving both multiple media, messages, and messengers, where one is not easy to distinguish from the other, as Marshall McLuhan, (1994) argued.

6 Conclusion

As planning processes in general, MSP is not just a technical exercise, one in which scientists have all the relevant knowledge. It is also important to sound out what small-scale fisheries people as the traditionally resident stakeholders know and aspire to. MSP is a process of transdisciplinary co-production. Access to what small-scale fisheries people know about the social and ecological system they are part of, is essential. But no less important is access to their 'phronesis'; the full array of their *habiti,* their values, ethics, worldviews, and language These are believed to be "sacred", as Berkes (1999) pointed out, and therefore not negotiable.

Planners must then know not just to argue convincingly but to listen respectfully, also to the subtext. In transdisciplinary MSP, the latter is as important as the former. The language of small-scale fisheries people is as metaphorical as that of scientists. They refer to a reality which scientists do not have direct access to. Neither have small-scale fisheries stakeholders to those of scientists. This creates the condition for "semantic infiltration", which would be detrimental to MSP as a process of transdisciplinary co-production.

Involving local people directly in the MSP process has inherent value. It is a matter of Blue Justice. Affected small-scale fisheries people have the right to be heard, as FAO (2015) points out. Giving phronetic knowledge a chance to inform the process also has functional merit. It creates a better, more informed, and legitimate plan. By involving small-scale fisheries people, planners will hear their criticism: how they think the system has flaws. They will also get their constructivism, their suggestions on how to fix the system. True co-production of MSP depends on it.

Conflicts may be a permanent challenge; they are not solved once and for all. There are conflicts about interests and values. MSP would, however, have made real progress if stakeholders could at least agree on the negative goals—what to avoid. Further marginalizing small-scale fisheries people is one of them. They have rights, also human rights, that MSP would do wrong to ignore. Compromise is perhaps the best thing one can hope for - a truce of sorts. It matters how conflicts are dealt with. People may accept outcomes that are less beneficial to them if they have been part of a process that is transparent and fair, where they had a say and are listened to. Presenting them with a *fait,* is a recipe for failure.

References

Acheson, J. A. (1998). *The lobster gangs of Maine.* The University Press of New England.
Bateson, G. (2000). *Steps to an ecology of mind. Collected essays in anthropology, psychiatry, evolution and epistemology.* Chicago University Press.

Berkes, F. (1999). *Sacred ecology. Traditional ecological knowledge and resource management*. Taylor and Francis.
Boudon, R. (1981). *The logic of social action. An introduction to sociological analysis*. Routledge & Kegan Paul.
Bourdieu, P. (1995). *Sociology in question*. Sage Publ.
Bourdieu, P. (2005). *The social structures of the economy*. Polity Press.
Buanes, A., & Jentoft, S. (2009). Building bridges. *Institutional Perspectives on Inter-disciplinarity. FUTURES., 41*(5), 446–454.
Caniglia, G., Freeth, R., Luederitz, C., et al. (2023). Practical wisdom and virtue ethics for knowledge co-production in sustainability science. *Nat Sustain, 6*, 493–501. https://doi.org/10.1038/s41893-022-01040-1
Chuenpagdee, R. and Jentoft, S. (Eds.) (2018). Transdisciplinarity for small-scale fisheries governance. *Analysis and Practice*. Springer Publ.
Das, J. (2023). Blue economy, blue growth, social equity, and small-scale fisheries: Global and National Level Review. *Studies. Social Science Research., 4, No. 1*. https://doi.org/10.22158/sssr.v4np38
Davis, T., Hanson, J. M., Watts, H., & MacPherson, H. (2004). Local ecological knowledge and marine fisheries research: The case of white hake (*Urophycis tenuis*) predation on juvenile American lobster (*Homarus americanus*). *Canadian Journal of Fisheries and Aquatic Sciences*. https://doi.org/10.1139/f04-070
Dyer, C. L., & McGoodwin, J. R. (Eds.). (1994). *Folk Management in the World's fisheries: Lessons for modern fisheries management*. University Press of Colorado.
Eskeland Schütz, S. and Slater, A-M. (2019). From strategic marine planning to project licenses—Striking a balance between predictability and adaptability in the management of aquaculture and offshore wind farms. *Marine Policy* Vol. 110, 103556.
FAO. (2015). *Voluntary guidelines for securing sustainable small-scale fisheries in the context of food security and poverty eradication*. UN Organization for Food and Agriculture.
FAO, WorldFish, and Duke University. (2023). *Illuminating Hidden Harvests*. https://www.worldfishcenter.org/hidden-harvests
Flannery, W. (2023). Making marine spatial planning matter. In: Partelow, S., Hadjimichael, M., Hornidge, AK. (eds) *Ocean Governance*. MARE Publication Series, vol 25. Springer. https://doi.org/10.1007/978-3-031-20740-2_5.
Flannery, W., Healy, N., & Luna, M. (2018). Exclusion and nonparticipation in marine spatial planning. *Marine Policy, 88*, 32–40.
Flyvbjerg, B. (2001). *Making social science matter. Why social inquiry fails and how it can succeed again*. Cambridge University Press.
Garland, M., Axon, S., Graziano, M., John Morrissey, J., & Heidkamp, C. P. (2019). *The blue economy: Identifying geographic concepts and sensitivities*. Geography Compass. https://doi.org/10.1111/gec3.12445
Goffman, E. (1986). *Frame analysis. An essay on the Organization of Experience*. Northeastern University Press.
Gudeman, S. (2001). *The anthropology of the economy. Community, market and culture*. Blackwell Publishing.
Habermas, J. (1984). Theory of Communicative Action, Volume One: Reason and the Rationalization of Society. Beacon Press. ISBN 978-0-8070-1507-0.
Haggan, N., Neiss. B. and Baird, I. G. (2007). Fishers' knowledge in fisheries science and management. Paris: Coastal management sourcebooks 4, UNESCO Publishing.
Hardin, G. (1968). The tragedy of the commons. *Science, 162*(3859), 1243–1248.
Heale, P. (1996). The communicative turn in planning theory and its implications for spatial strategy formation. *Environment and Planning B: Planning and Design., 23*(2), 217–234. https://doi.org/10.1068/b230217
Hilgers, M., & Mangez, E. (Eds.). (2015). *Bourdieu's theory of social fields. Concepts and applications*. Routledge.

Høst, J. (2015). *Market-based fisheries management. Private fish and captains of finance*. Springer Publ.
Janßen, H. Bastardie, F., Eero, M. et al. (2018). Integration of fisheries into marine spatial planning: Quo vadis? *Estuarine, Coastal and Shelf Science*. vol 201, pp 105–113.
Jentoft, S. (2006). Beyond fisheries management: The *Phronetic* dimension. *Marine Policy, 30*, 671–680.
Jentoft, S. (2023a). *The Gift of Community: More Essays on Human Experiences of Small-Scale Fisheries*. St. John's, Newfoundland: TBTI Global, 2023.
Jentoft, S. (2023b). The image fallacy; rethinking the tragedy of the commons. *International Journal of the Commons, 17*(1), 174–183. https://doi.org/10.5334/ijc.1248
Jentoft, S., Chuenpagdee, R., Isaacs, M., & Bugeja Said, A. (2022). *Blue justice. Small-scale fisheries in a Sustainable Ocean economy*. Springer Publ.
Johannes, R. E. (1981). *Words of the lagoon. Fishing and marine Lore in the Palay District of Micronesia*. University of California Press.
Jones, P. J. S., Lieberknecht, L. M., & Qiu, W. (2016). Marine spatial planning in reality: Introduction to case studies and discussion of findings. *Marine Policy, 71*, 256–264.
Kant, I. (1991). *The moral law*. Routledge Classics.
Keijser, X., Ripken, M., Mayer, I., et al. (2018). Stakeholder engagement in maritime spatial planning: The efficacy of a serious game approach. *Water, 10*, 724. https://doi.org/10.3390/w10060724
Lakoff, G., & Johnson, M. (1980). *Metaphors we live by*. The University of Chicago Press.
Lang D.J., Wiek, A., Bergmann, M., Stauffacher, M. et al. (2012). Transdisciplinary research in sustainability science: Practice, principles, and challenges. *Sustainability Science, 7* (Supplement 1):25–43. https://doi.org/10.1007/s11625-011-0149-x.
Malinowski, B. (1922/2010). Argonauts of the Western Pacific. An account of native Enterprise and adventure in the archipelagoes of Melanesian New Guinea. Routledge.
Maurstad, A. (2000). To fish or not to fish: Small-scale fishing and changing regulations of the cod fishery in northern Norway. *Human Organization, 59*(1), 37–47.
Mauss, M. (1954). *The gift: The form and reason for exchange in archaic societies*. W.W. Norton & Co.
McLuhan, M., & Lapham, L. H. (1994). *Understanding media: The extension of man*. The MIT Press.
Merton, R. K. (1948). *The Self-Fulfilling Prophecy. Antioch Review, 8*(2), 193–2010. https://doi.org/10.2307/4609267
Ostrom, E. (1990). *Governing the commons. The evolution of institutions for collective action*. Cambridge University Press.
Schutter, M. S., Hicks, C. C., Phelps, J., & Waterton, C. (2021). The blue economy as a boundary object for hegemony across scales. *Marine Policy, 132*, 104673.
Sennett, R. (2012). *Together. The rituals, pleasures, and politics of cooperation*. Yale University Press.
Smith, G. and Jentoft, S. (2017). Marine spatial planning in Scotland. Levelling the playing field? Marine Policy, vol 84, pp 33–41.
Sunde, J. (2014). Customary governance and expressions of living customary law at Deca-weber: Contributions to small-scale fisheries governance in South Africa, PhD Thesis. University of Cape Town.
Tafon, R. (2019). *The "dark side" of marine spatial planning. A study of domination, empowerment and freedom through theories of discourse and power*. Södertörn Doctoral Dissertations.
Thurow, L. C. (2001). *The zero-sum society. Distribution and the possibilities for change*. Basic Books.
Trouillet, B. (2019). Aligning with dominant interests. The role played by geotechnologies in the place given to fisheries in in marine spatial planning. *Geoforum, 107*, 54–65.

Visser, L. E. (2004). Reflections on Transdisciplinarity, integrated coastal development and governance. In L. E. Visser (Ed.), *Challenging coasts. Transdisciplinary excursions into integrated coastal zone development* (pp. 23–48). Amsterdam University Press.

Willmann, R., Franz, N., Fuentevilla, C., McInerney, T.F., and Westlund, L. (2017). Human Rights-Based Approach to Securing Small-Scale Fisheries: A Quest for Development as Freedom. (pp. 15–34) in Jentoft, S. et al. (eds.), *The Small-Scale Fisheries Guidelines. Global Implementation.* MARE Publication Series 14, https://doi.org/10.1007/978-3-319-55074-9_2.

Wittgenstein, L. (1953/2009). Philosophical investigations. West-Sussex: Blackwell Publ. Ltd.

Svein Jentoft is a sociologist and Professor Emeritus whose work focuses on marine resource management; small-scale fisheries; and coastal communities, both in the Global North and South. He is a founding member of the international network Too Big To Ignore (TBTI) and has led initiatives on governance of fisheries, particularly relating to FAO's SSF Guidelines (Voluntary Guidelines for Securing Sustainable Small-Scale Fisheries). His recent writing includes reflective and essay-style works about the human and community dimensions of small-scale fisheries, including Life Above Water and The Gift of Community.

Open Access This chapter is licensed under the terms of the Creative Commons Attribution-NonCommercial-NoDerivatives 4.0 International License (http://creativecommons.org/licenses/by-nc-nd/4.0/), which permits any noncommercial use, sharing, distribution and reproduction in any medium or format, as long as you give appropriate credit to the original author(s) and the source, provide a link to the Creative Commons license and indicate if you modified the licensed material. You do not have permission under this license to share adapted material derived from this chapter or parts of it.

The images or other third party material in this chapter are included in the chapter's Creative Commons license, unless indicated otherwise in a credit line to the material. If material is not included in the chapter's Creative Commons license and your intended use is not permitted by statutory regulation or exceeds the permitted use, you will need to obtain permission directly from the copyright holder.

Trade-Offs Conducting Action Research on the Commons: Balancing the Ethics of Research and Practice

Chris Giotitsas, Christina Priavolou, and Vasilis Kostakis

Abstract Action research is a methodological approach that imbues a research project with a normative goal and often involves the research subjects. This act may often occur in the context of a community, a localised social group, a school, or even a group of patients. Over the years, action research has acquired many variations, producing boundary-pushing research and striving to affect change in the various fields in which it has been applied. The goal of this chapter is twofold. First, through an empirical account of a research pilot that studies the commons, we argue that action research may benefit from a robust political economy underpinning. The latter may provide a structural element to the typically localised focus of action research. However, such an approach may accentuate the typical tensions between academic and activist work. So, this chapter also provides an account of the pains we experience, from a methodological perspective, to hopefully spark constructive dialogue on socially transformative and commons-oriented research.

1 Introduction

Over the years, action research (AR) has come to mean many different things in different iterations: from participatory action to community-based action, collective and collaborative efforts (Reason & Bradbury, 2006; Evered & Roger, 2022). With

C. Giotitsas · V. Kostakis
Ragnar Nurkse Department of Innovation and Governance, Tallinn University of Technology (TalTech), Tallinn, Estonia
e-mail: christos.giotitsas@taltech.ee

V. Kostakis
e-mail: vasileios.kostakis@taltech.ee

C. Priavolou (✉)
Amsterdam Institute for Social Science Research, University of Amsterdam, Amsterdam, Netherlands
e-mail: c.priavolou@uva.nl

V. Kostakis
Berkman Klein Center for Internet & Society, Harvard University, MA, USA

© The Author(s) 2026

S. Partelow (ed.), *Ethics and the Commons*, Schriften zu Leben und Gesellschaft - Series on Life and Society 3, https://doi.org/10.1007/978-3-032-24246-4_7

multiple roots in different scientific fields, which span back almost eight decades, the common denominator for the term is the act of imbuing a research project with a normative, if not socially transformative, goal and typically involves the research subjects in the process (McNiff & Whitehead, 2011; Rajbanshi & Luitel, 2020). This act usually takes place in the context of a community, a localised social group, a school, or even a group of patients.

Action research (AR) is particularly valuable when applied to investigate normative governance issues in the commons. The commons, characterised by shared resources managed collectively, require a deep understanding of community dynamics, resource management, and the intricate balance between individual rights and the collective good (Poonamallee et al., 2023). In such a context, where communities play a central role in resource management, involving community members in the research process is essential (Stringer et al., 2020). In addition, AR empowers communities to take an active role in managing common resources, leading to more effective governance structures, increased community cohesion, and a stronger sense of ownership over shared resources (Ambuehl et al., 2022). Also, AR projects in the commons can connect with other initiatives worldwide, contributing to a global network of shared knowledge and resources for the benefit of all.

Many theoretical approaches and methodological tactics, often with conflicting aspects, find application in action research (AR), making it challenging, if not futile, to categorise it within a specific academic discipline (Eden & Ackermann, 2018). This study departs from the conventional norm by presenting a vignette from our own research-activist pilot to address a recurring challenge in this domain. We contend that AR stands to gain immensely from a robust political economy underpinning, which provides a structural framework that extends beyond the typical localised focus by encompassing a profound comprehension of the broader political and economic forces shaping the research context. While a politically oriented AR can facilitate identifying and resolving systemic issues that may impede research success and transformative goals, it also intensifies the inherent tensions between academic rigor and activist imperatives (Cancian, 1993). Hence, from a methodological standpoint, this research note navigates the trade-offs we encounter when reconciling these contrasting elements, sparking constructive dialogue.

2 The Setting and the Challenge

Over the past decade, we have been actively engaged in AR within the framework of a ten-member research collective. What distinguishes our collective is its roots in political activism rather than a purely research-oriented foundation, underscoring our commitment to effecting social and political change. Our research endeavours serve as a means to advance our objectives, particularly those related to the commons and shared resources. This foundation influences our collaborative choices and shapes the nature of our research projects.

More specifically, a unifying theme throughout our work has been our focus on "the commons". In essence, we concentrate on social systems where goods are shared based on the rules and norms established by the productive community (as described by Bollier and Helfrich in 2019, as well as Kioupkiolis in 2019). We meticulously study these concepts and the corresponding fields of practice from two distinct angles: organisational structures and productive activities. In addition, our members hail from diverse disciplines, resulting in various research projects through which we advocate for political and social change, challenging established norms and structures and informing our approach to AR. Our collaborative efforts extend to students, coders/hackers, makers, farmers, builders, and other activists, making our AR initiatives a melting pot of expertise and passion.

Our latest project, which this chapter uses as a starting point, combines the various elements of previous AR projects to explore the potential of a production mode often called commons-based peer production or commoning (Bauwens et al., 2019). In such a mode, "the commons" embodies a profound shift in how we approach knowledge production and resource sharing. The essence of "the commons" lies in the global sharing of knowledge, designs, and software, veering away from traditional notions of tightly controlled information. This open-access approach allows communities worldwide to contribute, collaborate, and benefit from a wealth of shared resources. At the same time, this model localises manufacturing. This means that the production of physical goods aligns with local communities' unique needs and preferences. Community-driven facilities like makerspaces and fab labs empower individuals and communities by providing access to essential tools, equipment, and expertise, enabling them to produce goods actively. By doing so, we facilitate customisation and sustainability in the production process, promoting a more community-centric approach.

Our AR project involves three pilot locations, one in Greece, one in Tallinn, and another in Bhutan, but the aspiration is to increase them in the next five years (Kostakis et al., 2023). These locations serve as testbeds for commons-oriented technological production models, amalgamating various elements from our previous AR projects. Through these initiatives, we advocate for political and social change by challenging established norms and structures, all while shaping the future of AR within a commons-focused framework. The ultimate goal is to create a sustainable and adaptable production model that aligns with the principles of the commons.

The project's design reflects a commitment to technological innovation and community-building, with stakeholders playing a central role in shaping its direction and impact. The three ongoing pilots, active for three to five years, serve as experimental grounds for the project's main concept: utilising shared knowledge for localised manufacturing, guided by the principles of the commons. Manufacturing is done locally and adapted to local needs, harnessing the benefits of collective knowledge while ensuring that the resulting products or solutions are contextually appropriate and responsive to the unique requirements of each pilot location. In practical terms, we identify what technologies would be helpful in the area and try to develop them by utilising existing knowledge. On the other, we strengthen local bonds and cooperative practices with multiple stakeholders, including local community members, researchers, activists, makers, farmers, builders, and potentially local businesses and government entities who have an interest or involvement in the project's activities and outcomes. The project's effectiveness relies on its ability to navigate and foster collaboration among stakeholders across multiple levels of engagement, from the local community, business, government, and institutional levels to the global knowledge-sharing, policy, and education levels.

3 Negotiating Trade-Offs: Lessons from our Greek-Based Pilot

The Greek-based pilot called Tzoumakers is located in a rural municipality, which is also the place of origin and residence for several of the researchers involved and is composed primarily of farmers. The physical infrastructure for the pilot is a repurposed municipal building equipped with fabrication equipment (from 3D printers and Computer Numerically Controlled machines to metal cutters and screwdrivers). The local municipality played a vital role by offering this building and covering its associated fixed costs, such as electricity, water, and internet connectivity. That building also functions as a meeting space for the local community.

Given the Tzoumakers community's **limited resources and time** (most of them are small-scale farmers working in precarious conditions), we, as action researchers and members of the Tzoumakers community, organise "events" of various types. Meet-ups to discuss the next steps for the community and reflect on previous activity; prototyping workshops for new tools to be brainstormed and manufactured; exhibition workshops where people present their personal tools for feedback and further development; knowledge transfer workshops where certain members teach based on their skill sets (e.g., welding, woodworking, farming techniques, etc); gatherings to debate broader issues in the region.

In purely methodological terms, data collection is a mix of observation of activities and interactions within the community; participation in said activities; ongoing discussions with participants; and documentation of the productive output of the pilot (what types of tools and solutions emerge and why). Data analysis occurs in distinct streams whose primary focus varies between sustainability, organisational practices, and technological innovation.

As this is an AR project, the community is involved in all facets of the activity. Even the act of retrofitting the building was a collaborative effort. We also outfitted the space with the tools the community members deemed valuable. Many of the participants hold superior know-how regarding the technical elements of our activities compared to the members of the research collective. We have had various discussions with the local communities, communicating our vision and collectively deciding on issues like how we would operate as a community; what kind of tools we would create; or when we would meet. Overall, we have attempted to cultivate a sense of **co-ownership and autonomy in the community**, which would help it self-organise after our research project ends in 2024, meaning the funds to support it operationally. By fostering such strong community involvement and empowerment throughout the project, we aspire to leave a lasting legacy of sustainable self-governance, ensuring the project's sustainability beyond its completion, with continued community-led operational support.

AR typically provides the opportunity to investigate an issue/problem collectively. From collectively reflecting (and self-reflecting) on the research problem to better understand it; to deciding on a course of action; and ultimately implementing and refining the research process (McNiff & Whitehead, 2011). And there are a lot of things to consider regarding the research process. What is considered data in a complex project? How is it gathered and analysed? What is of relevance and to whom? Who ultimately benefits from the research results? So, when it comes to AR, it is rarely as clear-cut as a project description suggests. Instead, it is very case-specific, with the local specificities defining the best course of action. When the problem is clear and specific, like in education or healthcare applications, the answers to the above questions can be more or less straightforward.

In a project such as our own, however, it can be more **complex**. We are not only researching specific community issues. Indeed, the project's main goal is to study the community's practices collectively and determine ways to improve their situation. But together we explore something new as well. A way to organise and co-create differently inspired by other initiatives across the globe. The activity of these initiatives is widely diverse, corresponding to equally various problems. Hence, it is essential to strike a **balance between local needs and perspectives with global trends and practices**. Nevertheless, as hinted earlier, all these initiatives adhere to a specific political economy framework of making their productive output available for anyone to access using information and communication technologies. Thus, creating a pool of digital commons for these diverse initiatives to benefit from and contribute to is crucial.

Through this pilot, we build farming tools and a local community infrastructure that allows different types of relations to emerge and be sustained locally while simultaneously being part of a global network. For instance, connections with other farming communities, which largely inspired the pilot, or energy communities have been established through shared values and interests that lead to virtuous resource and knowledge exchange cycles.

4 Striking a Balance: Trade-Offs and Solutions

We believe that imbuing an AR project, which focuses on organisational and production practices, with a political economy element such as the one explained here, allows for scaling in social impact and research results in terms of generalisability. A thorny subject within AR often lies in its limited capacity to encompass broader themes beyond its apparent focus on local participation (Hickey & Mohan, 2005; Evans & Kotchetkova, 2009). To address this issue here, instead of relying on extrapolating local findings through diverse methodological approaches (see, for instance, Ponzoni, 2016; Brydon-Miller et al., 2003), we suggest generalisability and replicability be built into the methodological design of AR itself.

While general principles and models for AR already exist and serve as valuable guidelines (McNiff, 2013; Willis & Edwards, 2014), they can be customised and extended to address specific issues and objectives of individual research projects. In that sense, generalisability and replicability can be made into a de facto feature of the research driven by the underlying political economy. One prevalent approach to accomplish this is by employing a common framework (Partelow, 2023). The pilot becomes both hyper-local by focusing on the specificities of the local context and universal by imbuing it with the principles of openness and collaboration, which may connect it with like-minded initiatives across the globe (which in turn can also be studied for their own local insight). In other words, such an approach encourages the utilisation of knowledge resources and competencies developed elsewhere and the contribution to the broader pool of these digital commons. It also makes generalisability more organic in the local community without straining the partnership between researcher and participant, as it forms a central concept rather than a component that the researcher is attempting to incorporate as a perceived requirement of academic work.

In practical terms, the political economy framework guides decision-making processes, resource allocation, and project management, ensuring that the project remains aligned with both scholarly rigor and activist imperatives. In addition, prioritising community values and forming a sense of partnership ensures research is a meaningful and sustainable endeavour. Hence, an adaptive approach replaces the one-size-fits-all model with tailored research principles that naturally lead to generalisation through shared learning among diverse communities. Given the current predicament of our species, such provocative, adaptable research methods that challenge traditional notions of research validity in the pursuit of sustainable transformation are crucial (Bradbury & Waddell 2019). Consequently, through a comprehensive understanding of sociopolitical dynamics shaping the research context, researchers can enhance the transformative potential of participatory action research within the context of the commons-oriented production model.

Such an approach, however, does present complications and challenges. Both for the overall initiative itself and for us as researchers who attempt to develop and navigate a complex and material resource-dependent research project. Notably, we encounter dilemmas during community interactions, where our involvement might appear insincere. There are moments when our contributions could unintentionally steer the discussion to align with our political objectives rather than allowing the community's self-determined direction (Wittmayer & Schäpke, 2014). We have also observed that some participants may regard our opinions as having greater weight, given our substantial contributions to the project. As a result, despite our best efforts to position this as a community-driven endeavour, we grapple with the need to address our implicit biases and the explicit normative imperatives inherent to the project. This ethical dimension is paramount, often prompting a reflection on the underlying ethical frameworks that influence the community's development and the resulting research outputs.

More specifically, our approach aligns with the flexibility and adaptability often advocated by virtue ethics or social contractualism, rooted in practical wisdom and joint agreements. However, these local ethical approaches may contrast with broader, external ethical standards. For instance, our stance prioritises well-being and community feedback over rigid adherence to predetermined ethical frameworks (Banks et al., 2016; Bussu et al., 2020). These inconsistencies have been recognised, necessitating occasional deviations from conventional ethical mandates (Macleod et al., 2018; Temper, et al., 2019; Simon & Caroline, 2022; Wood & Kahts-Kramer, 2023).

Additionally, the complexities inherent in this approach may have implications for the potential research findings. Several inquiry streams operate at various levels, with intricate connections cutting across them all. These multifaceted streams can lead to diverse methods, approaches, and ethical commitments. For example, we are simultaneously interested in understanding individual behaviors and interpersonal interactions within this trans-local structure, as well as in fostering solid social structures within the local community characterised by equity and global collaboration. While these interests are not inherently contradictory, they give rise to moments of ethical dilemma and reflection.

Allow us to provide an example. During one of our more extensive workshops, groups were manufacturing different tools (Fig. 1). One of the most ambitious tools, a herb grinder, was planned and designed for a long time, and the team developing it was close-knit. Indeed, the time constraints inadvertently introduced challenges related to equity and collaboration. The trade-off here can be observed in the conflicting interests between those possessing advanced manufacturing skills and those who did not. As the manufacturing phase approached within a tight timeframe, individuals with the expertise understandably focused on the immediate task. Conversely, those without these skills found themselves on the periphery, unable to actively contribute and needing more time to acquire new skills through hands-on learning.

Some of us, wearing our researcher hats, noticed this dichotomy. We took notes regarding potential tensions created by expertise level discrepancies and continued our work in the workshop. But we did not attempt to bridge that divide at the time. We implicitly focused on allowing their independent interactions instead of more hands-on community building. At the end of the day, grievances emerged when we all assembled to discuss and reflect on the workshop. People felt left out. In hindsight, we regretted not stepping in earlier. However, this, as well as the discussion and resolution, was an instrumental and insightful development to study. So, negotiating priorities can be challenging and uneventful.

Another restriction more specific to this case is the material resource limitations that force choices between the feasible goals of the research project and the real aspirations for success both in terms of the community as well as the overarching production model. The tight time frame of the workshop above, which led to this tension, could have been better and something that was decided as the ideal course of action. But the truth is that the funds, along with the conditions that came with them and the time available, compelled us to do so. The extensive resources required

Fig. 1 Tools developed at Tzoumakers

(e.g., materials, equipment, and logistics costs) were only available in this format. If our situation allowed for different circumstances, we would choose to organise the workshop another way. We typically prefer to have plenty of time for manufacturing for everyone to be able to learn and participate.

Once again, we faced a pivotal choice that underlines the central tension between our research aspirations and the pragmatic requirements for the AR project to flourish as a genuine commoning initiative. The decision was clear: we opted to work with the resources at hand to keep the project on course and maintain the community's active participation. Pursuing an entirely organic approach would have posed formidable challenges, and finding a more suitable alternative appeared highly improbable. This situation underscores the intricate navigation of multiple roles, including those of a researcher, community member, political activist, and, perhaps most conspicuously, a project manager, which is a delicate balancing act. The following Fig. 2 summarises the contrasting aspects of research and AR outcome objectives, as discussed in this section, based on our experience in the three pilots.

Summing up, the political economy framework depicted in Fig. 2 offers a platform for the reconciliation of these contrasting features between researchers and activists. By delving into the intricacies of economic and political dynamics shaping the research context, this framework facilitates a collaborative approach to harness the strengths of both scholarly inquiry and grassroots activism towards shared goals of social justice. By identifying these dynamics, researchers and activists can align their goals toward addressing root causes related to systemic issues. Additionally, the political economy framework acknowledges the interconnectedness of social, economic, and political factors, enabling researchers and activists to adopt multifaceted strategies to drive change effectively. Lastly, emphasizing outcomes that drive tangible social and political transformation, insights from both scholarly research and activist praxis could be integrated into this framework to ensure meaningful impact on diverse stakeholders.

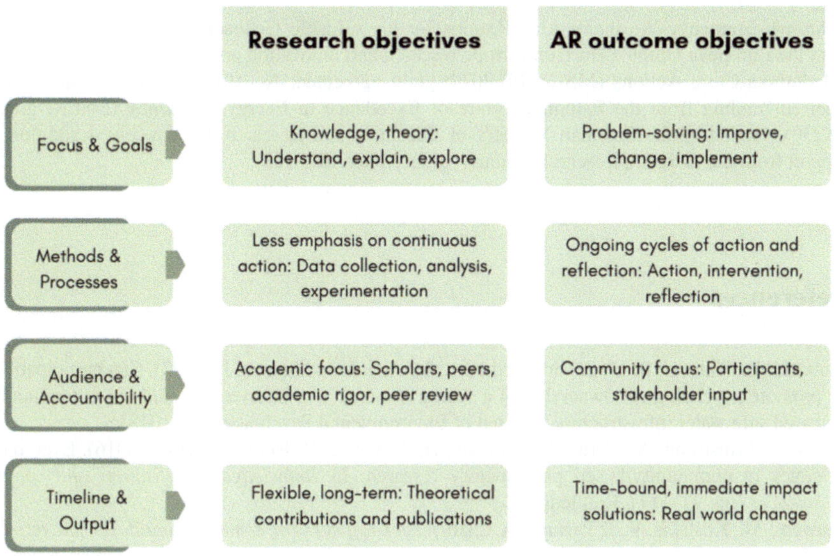

Fig. 2 Contrasting features of research and AR outcome objectives

5 Conclusion

We do not have an answer on how to cope with the typical tensions between academic and activist work. Attempting to build a framework to cover all contingencies would be an exercise in futility. One always needs to be mindful of the tensions and critical of their own role and intentions in a research endeavour. To always be honest in the struggle to find a balance between conducting a thorough and robust research project and trying to affect social change through, quite obviously, a political action. And then hope for the best. We find that it is worth the trouble.

This research approach is our way of securing the resources for our contribution to social transformation and supporting ourselves sufficiently while doing it- if occasionally precariously. AR, perhaps qualitative research in general to some degree, has always existed in an uneasy position within the overarching positivist atmosphere of rigorous academic research that still exists today. A balancing act between engaging in socially transformative, boundary-pushing research and conforming to cookie-cutter standards of assessing research proposals for funding ultimately limits the potential for this type of work to proliferate. The realities and structural obstacles of securing funds for such a project and later justifying its academic merit and "impact" escape the scope of this research note. Yet those who have engaged in such an effort may find it painful. It is now the time to push even harder for transformation in academia if it is to contribute to much-needed social change genuinely.

Acknowledgements We thank Alex Pazaitis for his valuable feedback. C.P. received funding from the European Union's Horizon Europe research and innovation programme under the Marie Skłodowska-Curie Actions (MSCA-PF-2024, grant agreement No 101206766). V.K. and C.G. received funding from the Estonian Centre of Excellence in Energy Efficiency (ENER; grant TK230), funded by the Estonian Ministry of Education and Research. C.G. received additional support from the Estonian Research Council (grant MOB3ERC123).

References

Ambuehl, B., Kunwar, B. M, Schertenleib, A., Marks, S. J. and Inauen, J. (2022). Can participation promote psychological ownership of a shared resource? An intervention study of community-based safe water infrastructure. Journal of Environmental Psychology 81: 101818.

Banks, S., Armstrong, A., Carter, K., Graham, H., Hayward, P., Henry, A., et al. (2016). Everyday ethics in community-based participatory research. In *Knowledge mobilisation and social sciences* (pp. 97–111). Routledge.

Bauwens, M., Kostakis, V., & Pazaitis, A. (2019). *Peer-to-peer: The commons manifesto*. University of Westminster Press.

Bollier, D., & Helfrich, S. (2019). *Free, fair, and alive: The insurgent power of the commons*. New Society Publishers.

Bradbury, H., Waddell, S., O' Brien, K., Apgar, M., Teehankee, B. and Fazey, I. (2019). A call to action research for transformations: The times demand it. Action Research 17(1): 3–10.

Brydon-Miller, M., Greenwood, D., & Maguire, P. (2003). Why action research? *Action Research, 1*(1), 9–28.

Bussu, S., Lalani, M., Pattison, S., & Marshall, M. (2020). Engaging with care: Ethical issues in participatory research. *Qualitative Research. Epub ahead of print*, (18 March 2020). https://doi.org/10.1177/1468794120904883

Cancian, F. M. (1993). Conflicts between activist research and academic success: Participatory research and alternative strategies. *The American Sociologist, 24*(1), 92–106.

Eden, C., & Ackermann, F. (2018). Theory into practice, practice to theory: Action research in method development. *European Journal of Operational Research, 271*(3), 1145–1155.

Evans, R., & Kotchetkova, I. (2009). Qualitative research and deliberative methods: Promise or peril? *Qualitative Research, 9*(5), 625–643.

Evered D and Roger I. S (2022) An assessment of the scientific merits of action research. An assessment of the scientific merits of action research, 135–161.

Hickey, S., & Mohan, G. (2005). *Participation: From tyranny to transformation? Exploring new approaches to participation in development*. Zed Books.

Kioupkiolis, A. (2019). *The common and counter-hegemonic politics: Re-thinking social change*. Edinburgh University Press.

Kostakis, V., Niaros, V., & Giotitsas, C. (2023). Beyond global versus local: Illuminating a cosmolocal framework for convivial technology development. *Sustainability Science, 18*, 2309–2322.

Macleod C.I, Marx J Mnyaka P and Treharne G.J (eds) (2018). The Palgrave handbook of ethics in critical research. Palgrave Macmillan.

McNiff, J. (2013). *Action research: Principles and practice* (3rd ed.). Routledge.
McNiff, J., & Whitehead, J. (2011). Action research. In *All you need to know about action research* (2nd ed.). Sage publications.
Partelow, S. (2023). What is a framework? Understanding their purpose, value, development and use. *Journal of Environmental Studies and Sciences, 13*(3), 510–519.
Ponzoni, E. (2016). Windows of understanding: Broadening access to knowledge production through participatory action research. *Qualitative Research, 16*(5), 557–574.
Poonamallee, L., Howard, A. D., & Joy, S. (Eds.). (2023). *Managing for social justice: Harnessing management theory and practice for collective good*. Springer Nature.
Rajbanshi R and Luitel B. C (2020) Transformative learning: An approach to understand participatory action research. Transformation 6(1): 2.
Reason, P., & Bradbury, H. (Eds.). (2006). *The sage handbook of action research. Participative inquiry and practice*. Sage.
Simon, W., & Caroline, S. (2022). Negotiating the ethical-political dimensions of research methods: A key competency in mixed methods, inter- and transdisciplinary, and co-production research. *Humanities & Social Sciences Communications, 9*(1), 1–13.
Stringer, T., Mortimer, G. and Payne, A. R. (2020). Do ethical concerns and personal values influence the purchase intention of fast-fashion clothing?. Journal of Fashion Marketing and Management 24(1): 99–120.
Temper, L., McGarry, D., & Weber, L. (2019). From academic to political rigour: Insights from the 'tarot ' of transgressive research. *Ecological Economics, 164*, 106379.
Willis, J. W., & Edwards, C. L. (2014). *Action research: Models, methods and examples*. Information Age Publishing Inc.
Wittmayer, J. M., & Schäpke, N. (2014). Action, research and participation: Roles of researchers in sustainability transitions. *Sustainability Science, 9*, 483–496.
Wood, L., & Kahts-Kramer, S. (2023). 'But how will you ensure the objectivity of the researcher?'guidelines to address possible misconceptions about the ethical imperatives of community-based research. *Research Ethics, 19*(1), 1–17.

Christos Giotitsas is a Senior Researcher at the Ragnar Nurkse Department of Innovation and Governance, Tallinn University of Technology (TalTech), Estonia. His research investigates commons-based peer production, degrowth, and sustainable innovation. He contributes to projects on digital transformation and localism in energy and technology governance.

Christina Priavolou is a Marie Skłodowska-Curie Postdoctoral Research Fellow at the University of Amsterdam, working on the intersection of sustainability, technology, and participatory design in the built environment. Her work also includes participatory action work with a rural makerspace in Greece, where she coordinated an EU-funded project on open and collaborative approaches to design and manufacturing.

Vasilis Kostakis is Professor of Technology Governance and Sustainability at TalTech and a Faculty Associate at Harvard University. His work combines theoretical research with practical action, co-building spaces for creative resistance and commons-based alternatives.

Open Access This chapter is licensed under the terms of the Creative Commons Attribution-NonCommercial-NoDerivatives 4.0 International License (http://creativecommons.org/licenses/by-nc-nd/4.0/), which permits any noncommercial use, sharing, distribution and reproduction in any medium or format, as long as you give appropriate credit to the original author(s) and the source, provide a link to the Creative Commons license and indicate if you modified the licensed material. You do not have permission under this license to share adapted material derived from this chapter or parts of it.

The images or other third party material in this chapter are included in the chapter's Creative Commons license, unless indicated otherwise in a credit line to the material. If material is not included in the chapter's Creative Commons license and your intended use is not permitted by statutory regulation or exceeds the permitted use, you will need to obtain permission directly from the copyright holder.

Navigating Ethical Spaces in Common Property Research

Graham Epstein, Rafaella Lobo, Evan J. Andrews, Maria B. Battaglia, Isabel Jorgensen, Ella Kari-Muhl, and Helena Shilomboleni

Abstract Common property research has long situated itself at the intersection of evidence-based research and advocacy that seeks to empower communities through local or collaborative governance of shared resources. While these identities often converge to provide a strong rationale for considering common property governance, they can sometimes diverge and raise important questions concerning our ability to satisfy our ethical obligations to local communities, external policymakers and the research community. Efforts to support community empowerment through close engagement in research may, for instance, bias the results and influence our ability to satisfy our ethical obligations to external stakeholders and the research community. In contrast, exclusion of communities can result in inaccurate or incomplete evidence about the commons that can serve to undermine our ethical obligations to all three groups and potentially contribute to the marginalization of communities. Although these tensions have long existed, they can no longer be ignored by common property researchers who must instead carefully examine the ethical implications of their approach to developing, undertaking and sharing the results of their

G. Epstein (✉) · M. B. Battaglia · I. Jorgensen · E. Kari-Muhl · H. Shilomboleni
School of Environment, Resources and Sustainability, University of Waterloo, Waterloo, Canada
e-mail: graham.epstein@uwaterloo.ca

M. B. Battaglia
e-mail: mbbattaglia@uwaterloo.ca

I. Jorgensen
e-mail: ibjorgen@uwaterloo.ca

E. Kari-Muhl
e-mail: ella-kari.muhl@uwaterloo.ca

H. Shilomboleni
e-mail: hshilomboleni@uwaterloo.ca

R. Lobo
Nicholas School of the Environment, Duke University, North Carolina, USA
e-mail: r.lobo@duke.edu

E. J. Andrews
Geography, Memorial University of Newfoundland, NL, Canada
e-mail: evana@mun.ca

© The Author(s) 2026

S. Partelow (ed.), *Ethics and the Commons*, Schriften zu Leben und Gesellschaft - Series on Life and Society 3, https://doi.org/10.1007/978-3-032-24246-4_8

research. In this chapter we draw upon our own experiences to identify consensus and contested standards for satisfying our ethical obligations to three core groups, including the communities that contribute to and inform our studies, the external stakeholders that may develop policies and programs on the basis of our research and the broader research community. In so doing we further identify areas of healthy dispute that illuminate our deep engagement with implementing best practices and good intentions.

1 Introduction

Research on the commons has developed rapidly over the last thirty years (van Laerhoven et al., 2020, Lambert et al., 2021) as the world continues to search for solutions to global environmental crises. Indeed, as evidence continues to mount concerning the relationship between community-based governance and sustainability (Chhatre & Agrawal, 2009, Cox et al., 2010, Persha et al., 2011, Cinner et al., 2016, Schleicher et al., 2017, Mawa et al., 2021) common property governance has emerged as an important element of global strategies to address these crises (Gurney et al., 2021). At the same time, other studies have advocated for the adoption of common property governance on the basis of normative principles of community empowerment, equity, justice, and reconciliation (Bennett et al., 2018, Blythe et al., 2023), independent of the environmental outcomes they generate. While the combination of empirical evidence and normative justifications for common property governance offer a range of compelling reasons for considering community-based solutions, it also raises important questions concerning the ability of researchers to satisfy their ethical obligations to three core groups of actors. These include the rightsholders and communities that use and manage the commons, external stakeholders working in governmental and non-governmental organizations, and the research community. In this chapter we therefore reflect on our own experiences in grappling with the ethics of our research and how we can satisfy our ethical obligations to these three groups.

The authors of this chapter share a common goal in seeking to ensure that our research avoids harms and generates knowledge to support advances in theory, sustainability and the social and economic wellbeing of communities. However, we differ in terms of the ethical principles and priorities that inform our approach to research and the pathways by which we seek to support advances in theory and practice. These perspectives have been shaped by our personal values, experiences growing up in North America, South America, Europe and Africa and graduate training in Canada and the United States. This has led some of us to give priority towards our ethical obligations to the research community and generating consistent and reliable evidence to inform both theory and practice. In contrast, experiences have led others to give priority to our ethical obligations towards communities in seeking to ensure that our research is aligned with their interests, needs and values. While these ethical positions can and do align in some cases, tensions will often emerge in field research. As a result, this chapter aims to help scholars navigate these tensions by drawing upon the diversity of positions among our team to identify consensus standards for ethical research on the commons. In so doing, we acknowledge that there were many areas where disagreements remained (i.e., contested standards), particularly regarding our ethical obligations to communities. These areas of disagreement may merit special attention from commons researchers before collaborating on field research to ensure that research activities are aligned with one's core ethical perspectives.

The remainder of the paper is structured in the following way. Section 2 begins by first elaborating upon the high-level ethical principles or values that underlie our respective positions and respective approaches to ethical research. Section 3, meanwhile, continues by discussing the ways in which these principles can be implemented to satisfy our ethical obligations to local communities, external stakeholders and researchers.

2 Ethical Principles

Ethical research on the commons begins first and foremost with a long-term commitment to reflect on one's own role in the research process and how this may directly or indirectly influence the results and the wellbeing of communities and broader society. Those operating from a more classical (and Western) approach to research prioritize our ethical obligations to the research community and an aspirational (and unachievable) goal of generating unbiased knowledge about the governance of the commons. This process generally begins with a careful consideration of potential sources of bias and the development of strategies to minimize their impact on findings. Focus groups and key informant interviews involving local leaders may, for instance, systematically bias responses towards the knowledge and experiences of powerful individuals within communities (Morgan, 1997, Fleming et al., 2022, Mahajan et al., 2023). These individuals may not, however, be the most knowledgeable when it comes to the governance of common-pool resources, and in some cases may even have incentives to misrepresent the status of local governance systems. As a result, although it may be necessary and beneficial to engage with local leaders in research, ethical researchers can and should take steps to better represent the diversity of community perspectives or at least be transparent about the limitations of their methods. For example, the Wildlife Conservation Society's coral reef social-ecological monitoring program combines key informant interviews with household surveys to offer a more balanced perspective of the structure and functioning of local governance systems (Gurney et al., 2019).

The emphasis on our ethical obligations to the research community does not however, imply a lack of obligation towards communities and external stakeholders. Indeed efforts to be transparent about our methods and potential sources of bias are one way that we try to build trust in our research. This in turn can help to increase confidence among external stakeholders that may use our research to inform the design of policies and programs that benefit communities. Furthermore, giving priority to serving the research community does not preclude close engagement with communities and external stakeholders in research. Engagement may, for instance, help to align research with local needs and priorities and support the development of a better understanding of the local context and processes. However, for those of us that give priority to more classical evidence-based research, close engagement is not always necessary for ethical research.

In contrast, other members of our team of co-authors view close collaboration with communities as a foundational element of ethical research. We believe that research can never be value-free (Haraway, 1988), and that often well-intended actors can perpetuate or exacerbate inequalities and power relations (Orr & Bennett, 2009; Reed et al., 2020; Mochizuki & Wada, 2023). Research is not just an objective assessment of "how the world works," but it is also a reflection of the asymmetrical power relations which place the researcher in a position of power. Ethics in research from this point of view means constant reflexivity and reflections on positionality. As a result, if our starting point is that we and our research are embedded in systems of power relations, then the challenge becomes how we manage our own entanglements with those systems to dismantle hierarchies and distance from communities (Sundberg, 2015). This in turn requires researchers to examine their motivations and methods to confront their own "entanglement ethics" by asking and continually reflecting upon the questions outlined in Box 1.

Box 1: Questions of reflection for entanglement ethics

1. Whose interests are being served by this research? (Nayak and Berkes, 2010)
2. Is our decision-making influenced by theoretical questions, personal values and perspectives or opportunities for career advancement and development? (Sundberg, 2015)
3. To what extent is our research extractive in nature? (Wainwright, 2012);
4. Does our research treat other human beings as subjects, objects, or data points? And how does this perspective enable or restrict different possibilities?
5. Are we supporting local communities or could our results be used to establish systems that restrict local resource users access to natural resources that meet their basic needs? (Joshi, 2015, Sundberg, 2015; Partelow et al., 2023)
6. Are we employing analytical categories that serve to reproduce difference and prejudice, further silencing the voices of those we wish to understand? (Chakrabarty, 2000, Sundberg 2015)
7. How do our analytical choices silence or empower certain voices/peoples over others? (Fleming et al., 2022)
8. How does our own positionality influence or inform the research we choose to pursue? (Haraway, 1988)

These are just some of the many ethical questions reflexivity requires. The point is not to feel overwhelmed with guilt or burdens, but rather to never lose sight of your own positionality and personal ethics, while engaging with people and institutions that are situated in different power positions.

3 Ethical Research Standards

In this section we draw upon the experiences, values and perspectives of our team of authors to synthesize recommendations for ethical research on the commons as it relates our ethical obligations to communities, external stakeholders and the research community. More specifically, we used Table 1 to first list and then organize our own perspectives regarding our ethical obligations to these groups. Consensus standards are those for which there was unanimous agreement among authors, while disputed standards are those for which there was at least some level of disagreement among the authors of this paper. Although unanimous agreement might seem a strong requirement for establishing standards, some ethical principles are formed from the core beliefs and values of individuals and cannot be resolved by debate or negotiation.

Additional considerations are principles that may further help to satisfy our ethical obligations to a particular group but are not absolutely required and may involve trade-offs or risks with respect to other ethical obligations. For example, while data sharing with other researchers is generally desirable (Tanhua et al., 2019, Alexander et al., 2020), disclosure of certain information about individuals, groups or communities may generate or exacerbate conflicts, cause harm and contribute to the further marginalization of some groups (Ross et al., 2018). As a result, our team generally agreed that one might consider sharing data with external stakeholders and the research community, but none of us felt that data sharing was necessary to satisfy our ethical obligations to these groups. In contrast, there was some dispute among our team regarding the need to share data with communities. While some feel that data sharing is essential for ethical research, others had concerns that some types of data sharing might generate or exacerbate conflicts within communities. Section 3 therefore continues by reviewing and discussing the principles outlined in Table 1.

3.1 Ethical Obligations to Communities

Communities and the rightsholders that use and manage common-pool resources are a cornerstone of common property research. They have shared rich empirical insights concerning the diversity of ways in which communities have developed and adapted institutions to sustainably govern the use of resources (Johannes, 1978, Ostrom, 1990, Wade, 1994, Nansikombi et al., 2020), that constitute the foundation of common property theory. They are often the primary subjects of our research as we seek to develop a better understanding of the social, ecological, and institutional factors that influence collective action, rule compliance, and sustainable governance of resources. However, communities are increasingly playing an active role in the design, implementation, and interpretation of research. This trend has emerged in part as a response to growing critiques of 'parachute science' (Asase et al., 2022, de Vos & Schwartz, 2022) where external researchers arrive to extract data and provide little to

Table 1 Proposed standards and recommendations for ethical research

	Obligations to local communities	Obligations to external stakeholders	Obligations to researchers
Consensus standards	• Obtain required research ethics approvals. • Obtain free prior and informed consent from research participants. • Design research to account for power dynamics within and between communities • Consider barriers to participation and strategies to address them. • Share research findings in diverse accessible formats. • Be transparent in terms of the role of the community in the research. • Develop understanding concerning access to data and ownership. • Never over-promise and establish clear expectations regarding potential benefits. • Critically evaluate outputs to consider the potential subjective interpretations of external stakeholders • Carefully select research partners inside and outside communities to ensure adherence to ethical practices. • Share affiliations and source of research funding	• Develop understanding concerning access to data and ownership. • Share research findings. • Provide guidance concerning the limitations of the research. • Provide guidance concerning how findings can be used to inform activities. • Provide guidance concerning the transferability of findings. • Prepare an executive summary. • Share affiliations and source of research funding.	• Share research findings. • Be transparent in terms of data collection, methods, and limitations. • Provide guidance concerning the transferability of findings. • Share affiliations and source of research funding.
Contested standards	• Share data with communities. • Ensure and support community data sovereignty. • Member checking (community validation of data) • Obtain community approval for research outputs. • Co-develop knowledge mobilization plans. • Develop a memorandum of understanding regarding outputs and next steps following the project end date to manage expectations		
Additional considerations	• Seek input from community leaders and members regarding research ethics. • Advocacy and support for communities. • Support community empowerment and emancipation • Consider the role of the researcher in the community (i.e., positionality) • Co-create research projects in partnership with rights holders and other affected stakeholders.	• Share data with external practitioners. • Co-develop research	• Share data with researchers. • Share step-by-step details on methods for data collection and analysis.

nothing in return. While there is unanimous agreement among our team that communities' merit special attention when it comes to the ethics of our research, some of us question whether community participation is always essential for ethical research. Community members can and do bring a wealth of local knowledge that allows us to tailor research to local contexts and contextualize our observations of complex social and ecological phenomena. However, community participation in research can also serve to bias the results through knowledge of research questions and methods, and in some cases can serve to reinforce local inequalities if participation is limited to rent-seeking local elites.

Among our team of co-authors there are diverse perspectives regarding our ethical obligations to communities. At one end there are those of us that believe we can largely satisfy our ethical obligations to communities by: (i) obtaining and adhering to required ethics approvals, (ii) being transparent in terms of the purpose and potential benefits of our research, (iii) seeking free, prior and informed consent from participants, and (iv) sharing our research findings with those communities in a variety of accessible formats. In contrast, some of us believe that while these principles are important and necessary, they are inadequate for ethical research. In fact, some Indigenous scholars view research ethics boards as a tool to protect universities from lawsuits, not local communities from harm (e.g. Louis, 2007). It also leaves unquestioned the "right of the researcher to pursue knowledge about whatever, wherever, whenever, and with whomever she chooses" (Sundberg, 2015).

Although we view science as a valuable tool for the construction of a better world, we believe that it is always a situated perspective. And even if the research can be considered "neutral" or value-free, it may be used in ways that are not. As such, we share what Bryant and Jarosz (2004, p. 807) called a "basic radical ethical position". Simply put, our position is one that privileges the rights and concerns of the communities we study, throughout the whole research process. For instance, if we share the results at the end of the project, and community members object to it, then that research is not done. If halfway through the research, the community decides they no longer want us there, then we need to pack up and leave. While we always hope and do our best to maintain our commitments and responsibilities to all stakeholders, if there is ever a conflict of interest, our ethical commitment to the research subjects supersedes that to any other stakeholder, including ourselves.

Despite our differences, there was a general agreement that there is a need for co-developed or co-produced research to enable alignment between research with interests, needs and values of local communities. Knowledge co-production generally refers to the collaborative process of bringing a plurality of knowledge sources and types together to formulate and address a defined problem to build a systems-oriented understanding of that problem for an actionable outcome (Cornwall, 2008; Armitage et al., 2011; Norström et al., 2020; Muhl et al., 2023). This could also include discussions regarding the ethics of proposed research methods and instruments. However, we failed to reach a consensus regarding the extent to which co-production is essential for field research on the commons. While some of us believe that co-production should be a standard for field research, others view co-production as one part of a larger set of research approaches broadly including co-produced, community-led and externally driven research. For us, the diversity of approaches offers a potentially powerful approach for triangulating results and offsetting the biases associated with each individual approach. However, some of us feel that the goal of generalizability is itself an implicit bias shared by positivist scientists, one that is not harmless or value-free. For us, we believe it is our ethical responsibility to be very careful about how these simplifications "hide complexities and alternatives" and "lend themselves to claims of truth" (Carpenter, 2020).

3.2 Ethical Obligations to External Stakeholders

External stakeholders are the individuals or groups responsible for a range of activities relevant to commons governance, including the development of laws and policies, monitoring, information sharing, capacity building and allocation of rights and financial resources. They include politicians, government agencies, international organizations like the International Union for Conservation of Nature, non-governmental organizations like the World Wildlife Fund, donors, and in some cases private enterprises responsible for the development of natural resource, conservation policies, and programs. As can be seen in Table 1, we generally agreed that we have a smaller number of ethical obligations to external stakeholders when compared to communities. Nonetheless, as actors with the authority, resources and capacity to develop policies and programs on the basis of our research, our direct and indirect interactions with this group have potentially wide-ranging ethical implications. Research studies that overstate results or neglect discussing limitations that may influence findings, risk having wide-reaching impacts on the environment and well-being of communities (Partelow et al., 2023). The Tragedy of the Commons, for instance, developed global recommendations for the governance of the commons based upon an allegorical model with fundamentally erroneous assumptions about the nature of human motivation and decision-making (Ostrom, 1998, 2007) and resulted in the dismantling of effective common property systems around the world (Johannes, 1978, Dobson & Lynch, 2003).

There was a high level of consensus regarding our ethical obligations to external stakeholders. In general, we agree that researchers can and should make efforts to share research findings with external stakeholders beyond traditional academic publications and presentations. These could include high-level executive summaries that highlight the major findings of our research, its limitations and provide guidance concerning how the results could be used to inform the design of policies, programs or activities. In particular as external stakeholders often operate at large geographical scales it is generally important to avoid over-extrapolation of results and discuss factors that might influence the transferability of findings to other contexts and communities. In fact, one of the important ways that we can help to satisfy our ethical obligations to communities is considering how our work will be interpreted and operationalized as part of a discourse or policy by external stakeholders.

Researchers may also consider sharing data with external stakeholders, enabling them to explore other questions, contextualize the results and compare findings across different communities and contexts. However, we generally concur that data sharing is not necessary to satisfy our ethical obligations to external stakeholders. Instead, researchers must first ensure that data sharing does not violate ethical obligations established with research participants who initially shared the data, and second consider carefully the extent to which data sharing could result in harms to individuals or communities. Furthermore, any data that is shared should always be de-identified unless the individual explicitly wants to be named or is identified as playing a key role and has given permission to be identified in the study (e.g., in a leadership position). Finally, researchers might consider co-developing research with external partners like governments or NGOs to help enhance the policy relevance of research. In this situation, ethical obligations may need to be discussed, especially in decision-making settings where no formalized research ethics process exists related to the types of research about to be undertaken. Then, this partnership should be undertaken carefully to ensure that it does not undermine our ethical obligations to communities, and clearly discuss how the participation of external stakeholders may have influenced the results.

3.3 Ethical Obligations to the Research Community

Third and finally the research community are individuals and groups engaged in theoretical and applied research on the commons. They include students, faculty, and postdoctoral researchers working at universities seeking to support advances in commons theory, as well as applied researchers and consultants working to support planning, monitoring, evaluation, and learning for a range of different public, private and non-profit organizations. Indeed, although social science capacity generally lags natural science capacity within public and non-profit conservation organizations (Bennett et al., 2016, 2017), there are a growing number of social scientists working to support the transition towards more effective, equitable, and just environmental governance. This in turn raises further important ethical considerations concerning our affiliations and sources of funding and ensuring this does not have an undue influence on our research or its findings. At a minimum, researchers should disclose their affiliations (including the country in which their organizations are located if conducting research across national borders) and sources of funding as part of an informed consent process for members of communities and in any materials, reports or presentations resulting from the research.

Our perspectives with regards to our ethical obligations to the research community were broadly similar to our perspectives on external stakeholders. We all agreed that it is important that we share our findings with the research community through presentations and publications so that other researchers can learn from our work, and potentially leverage those findings to uncover patterns across a larger number of studies (i.e. Ostrom, 1990, Cox et al., 2010, Baggio et al., 2016). At the same time, it is also important that we are transparent in terms of how data was collected and analyzed, and clearly describe the core limitations of our research. In general, descriptions of methods for data collection and analysis, and their associated limitations should be more detailed in outputs directed at the research community. This would allow them to draw upon their own skills and knowledge to evaluate the strengths and weaknesses of the research and apply the same approach to collect comparable data. At a minimum it is essential that we are honest about our research methods and avoid the temptation to overstate the strength of our methods, analysis or findings and instead be clear regarding its limitations. For example, the concept of co-production with communities is often invoked by researchers, despite limited or sometimes token engagement with communities (Cornwall, 2008). Similarly, problems with sampling, instruments, indicators, and the timing of research activities can alter the final outcomes. Lessons regarding spatial, temporal, and other constraints should be shared so that others can better understand the limitations of the results, and potentially develop strategies for addressing them in their own research. As a result, careful reflection of research activities as planned and as implemented provide an important foundation for satisfying our ethical obligations to the research community by supporting the development of clear and honest descriptions of our methods, results, and associated challenges.

4 Conclusions

Standards and principles for ethical research on the commons have evolved over the last thirty years. Whereas many of the foundational case studies that informed Ostrom's (1990) design principles would have relied exclusively upon the ethical principles of individual researchers, current research is increasingly guided by standards established by research ethics boards and committees. However, beyond a few requirements, such as seeking free, prior and informed consent, individual researchers continue to exercise considerable discretion in terms of their approach to ethical research. This paper, which synthesizes the perspectives of researchers with different life experiences, training and values that share a common interest in common property governance may help researchers to grapple with the ethics of their research by:

1. Providing an expanded set of ethical obligations to communities, external stakeholders and the research community beyond those established by research ethics boards and committees.
2. Highlighting ethical obligations to communities as a special class of ethical obligations that require detailed attention and reflection and serve to both unite and divide common property researchers.
3. Revealing key areas of contestation over ethical obligations (i.e., data sharing with communities, community approval of research outputs) that may generate tensions among different researchers, communities, and stakeholders. These

areas of contestation should be discussed early in the research process, including potential trade-offs and their implications to ensure alignment across members of a research team

4. Be transparent when reporting findings to communities, external stakeholders and the research community.

References

Alexander, S. M., Jones, K., Bennett, N. J., Budden, A., Cox, M., Crosas, M., Game, E. T., Geary, J., Hardy, R. D., & Johnson, J. T. (2020). Qualitative data sharing and synthesis for sustainability science. *Nature Sustainability, 3*, 81–88.

Armitage, D., Berkes, F., Dale, A., Kocho-Schellenberg, E., & Patton, E. (2011). Co-management and the co-production of knowledge: Learning to adapt in Canada's Arctic. *Global Environmental Change, 21*(3), 995–1004.

Asase, A., Mzumara-Gawa, T. I., Owino, J. O., Peterson, A. T., & Saupe, E. (2022). Replacing "parachute science" with "global science" in ecology and conservation biology. *Conservation Science and Practice, 4*, e517.

Baggio, J., Barnett, A., Perez-Ibarra, I., Brady, U., Ratajczyk, E., Rollins, N., Rubiños, C., Shin, H., Yu, D., & Aggarwal, R. (2016). Explaining success and failure in the commons: The configural nature of Ostrom's institutional design principles. *International Journal of the Commons, 10.*

Bennett, N. J., Kaplan-Hallam, M., Augustine, G., Ban, N., Belhabib, D., Brueckner-Irwin, I., Charles, A., Couture, J., Eger, S., & Fanning, L. (2018). Coastal and indigenous community access to marine resources and the ocean: A policy imperative for Canada. *Marine Policy, 87*, 186–193.

Bennett, N. J., Roth, R., Klain, S. C., Chan, K., Christie, P., Clark, D. A., Cullman, G., Curran, D., Durbin, T. J., Epstein, G., Greenberg, A., Nelson, M. P., Sandlos, J., Stedman, R., Teel, T. L., Thomas, R., Veríssimo, D., & Wyborn, C. (2017). Conservation social science: Understanding and integrating human dimensions to improve conservation. *Biological Conservation, 205*, 93–108.

Bennett, N. J., Roth, R., Klain, S. C., Chan, K. M. A., Clark, D. A., Cullman, G., Epstein, G., Nelson, M. P., Stedman, R., Teel, T. L., Thomas, R. E. W., Wyborn, C., Curran, D., Greenberg, A., Sandlos, J., & Veríssimo, D. (2016). Mainstreaming the social sciences in conservation. *Conservation Biology, 31*, 56–66.

Blythe, J. L., Gill, D. A., Claudet, J., Bennett, N. J., Gurney, G. G., Baggio, J. A., Ban, N. C., Bernard, M. L., Brun, V., Darling, E. S., Di Franco, A., Epstein, G., Franks, P., Horan, R., Jupiter, S. D., Lau, J., Lazzari, N., Mahajan, S. L., Mangubhai, S., Naggea, J., Turner, R. A., & Zafra-Calvo, N. (2023). Blue justice: A review of emerging scholarship and resistance movements. *Cambridge Prisms: Coastal Futures, 1*, e15.

Bryant, R. L., & Jarosz, L. (2004). Ethics in political ecology: A special issue of political geography introduction: Thinking about ethics in political ecology. *Political Geography, 23*(7), 807–812.

Carpenter, C. (2020). *Power in conservation: Environmental anthropology beyond political ecology.* NY, Routledge.

Chakrabarty, D. 2000. A small history of subaltern studies. A companion to postcolonial studies:467–485.

Chhatre, A., & Agrawal, A. (2009). Trade-offs and synergies between carbon storage and livelihood benefits from forest commons. *Proceedings of the National Academy of Sciences, 106*, 17667–17670.

Cinner, J. E., Huchery, C., MacNeil, M. A., Graham, N. A., McClanahan, T. R., Maina, J., Maire, E., Kittinger, J. N., Hicks, C. C., & Mora, C. (2016). Bright spots among the world's coral reefs. *Nature.*

Cornwall, A. (2008). Unpacking 'participation': Models, meanings and practices. *Community Development Journal, 43*(3), 269–283.

Cox, M., Arnold, G., & Villamayor Tomas, S. (2010). A review of design principles for community-based natural resource management. *Ecology and Society, 15*, 38.

Dobson, T. A., & Lynch, K. D. (2003). As nearshore stocks drop, Malawi begins a return to local fisheries management. *Journal of Great Lakes Research, 29*, 232–242.

Fleming, W., King, B., Robinson, K., Wade, E., Erickson, B., Delie, J., de Ycaza, R., Trimbach, D., Spalding, A., & Biedenweg, K. (2022). Is there no "I" in team? Potential bias in key informant interviews when asking individuals to represent a collective perspective. *PLoS One, 17*, e0261452.

Gurney, G. G., E. S. Darling, G. N. Ahmadia, V. N. Agostini, N. C. Ban, J. Blythe, J. Claudet, G. Epstein, A. Himes-Cornell, and H. D. Jonas. 2021. Biodiversity needs every tool in the box: Use OECMs. Nature Publishing Group.

Gurney, G. G., Darling, E. S., Jupiter, S. D., Mangubhai, S., McClanahan, T. R., Lestari, P., Pardede, S., Campbell, S. J., Fox, M., & Naisilisili, W. (2019). Implementing a social-ecological systems framework for conservation monitoring: Lessons from a multi-country coral reef program. *Biological Conservation, 240*, 108298.

Haraway, D. (1988). Situated knowledges: The science question in feminism and the privilege of partial perspectives. *Feminist Studies, 14*(3), 575–599.

Johannes, R. E. (1978). Traditional marine conservation methods in Oceania and their demise. *Annual Review of Ecology and Systematics, 9*, 349–364.

Joshi, S. 2015. Postcoloniality and the North–South binary revisited: The case of India's climate politics. International Handbook of Political Ecology:117–130.

van Laerhoven, F., M. Schoon, and S. Villamayor-Tomas. 2020. Celebrating the 30 th Anniversary of Ostrom's Governing the Commons: Traditions and Trends in the Study of the Commons, Revisited. International Journal of the Commons 14.

Lambert, J., Epstein, G., Joel, J., & Baggio, J. (2021). Identifying topics and trends in the study of common-Pool resources using natural language processing. *International Journal of the Commons, 15.*

Louis, R. (2007). Can you hear us now? Voices from the margin: Using indigenous methodologies in geographic research. *Geographical Research, 45*(2), 130–139.

Mahajan, S. L., S. Obiene, L. Ojwang, N. Olwero, A. Valdivia, A. Wosu, E. Adrid, D. A. Andradi-Brown, G. Andriamalala, N. C. Ban, N. J. Bennett, J. Blythe, S. H. Cheng, E. Darling, M. De Nardo, E. DruryO'Neill, G. Epstein, R. Y. Fidler, K. Fisher, J. Geldmann, D. A. Gill, R. G. Kroner, G. Gurney, A. Jagadish, H. D. Jonas, M. E. Lazuardi, S. Petersen, V. V. Ranarivelo, L. Rasoloformanana, T. M. Rasolozaka, D. J. Read, E. S. Mwaiteleke, and G. Ahmadia. 2023. Introducing Elinor for monitoring the governance and management of area-based conservation. Conservation Biology n/a.

Mawa, C., Tumusiime, D. M., & Babweteera, F. (2021). Are community forests delivering livelihood benefits? Insights from Uganda. *Forests, Trees and Livelihoods, 30*, 133–150.

Mochizuki, J., & Wada, Y. (2023). Closing the loop of reflexivity. *Nature Climate Change, 13*, 110–112.

Morgan, D. L. (1997). *Focus groups as qualitative research* (2nd ed.). Sage Publications.

Muhl, E. K., Armitage, D., Anderson, K., Boyko, C., Busilacchi, S., Butler, J., Cvitanovic, C., Faulkner, L. A., Hall, J. A., Martynuik, G., & Paul-Burke, K. (2023). Transitioning toward "deep" knowledge co-production in coastal and marine systems: Examining the interplay among governance, power, and knowledge. *Ecology and Society, 28*(4).

Nansikombi, H., Fischer, R., Kabwe, G., & Günter, S. (2020). Exploring patterns of forest governance quality: Insights from forest frontier communities in Zambia's miombo ecoregion. *Land Use Policy, 99*, 104866.

Nayak, P. K., & Berkes, F. (2010). Whose marginalisation? Politics around environmental injustices in India's Chilika lagoon. *Local Environment, 15*, 553–567.
Norström, A. V., Cvitanovic, C., Löf, M. F., West, S., Wyborn, C., Balvanera, P., Bednarek, A. T., Bennett, E. M., Biggs, R., de Bremond, A., & Campbell, B. M. (2020). Principles for knowledge co-production in sustainability research. *Nature Sustainability, 3*(3), 182–190.
Orr, K., & Bennett, M. (2009). Reflexivity in the co-production of academic-practitioner research. *Qualitative Research in Organizations and Management, 4*(1), 85–102.
Ostrom, E. (1990). *Governing the commons*. Cambridge University Press.
Ostrom, E. 1998. A behavioral approach to the rational choice theory of collective action: Presidential address, American Political Science Association, 1997. The American Political Science Review 92:1–22.
Ostrom, E. (2007). A diagnostic approach for going beyond panaceas. *Proceedings of the National Academy of Sciences of the United States of America, 104*, 15181–15187.
Partelow, S., Schlüter, A., Ban, N. C., Batterbury, S., Bavinck, M., Bennett, N. J., Bleischwitz, R., Blythe, J., Bogusz, T., Breckwoldt, A., Cinner, J. E., Glaser, M., Govan, H., Gruby, R., Hatje, V., Hornidge, A.-K., Hovelsrud, G. K., Kittinger, J. N., Kluger, L. C., Kochalski, S., Mawyer, A., McKinley, E., Olsen, J., Pittman, J., Riechers, M., Riekhof, M.-C., Manez, K. S., Shellock, R. J., Siriwardane-de Zoysa, R., Steins, N. A., Van Assche, K., & Villasante, S. (2023). Five social science intervention areas for ocean sustainability initiatives. *npj Ocean Sustainability, 2*, 24.
Persha, L., Agrawal, A., & Chhatre, A. (2011). Social and ecological synergy: Local rulemaking, Forest livelihoods, and biodiversity conservation. *Science, 331*, 1606–1608.
Reed, G., Dagli, W., & Hambly Odame, H. (2020). Co-production of knowledge for sustainability: An application of reflective practice in doctoral studies. *Reflective Practice, 21*(2), 222–236.
Ross, M. W., Iguchi, M. Y., & Panicker, S. (2018). Ethical aspects of data sharing and research participant protections. *American Psychologist, 73*, 138.
Schleicher, J., Peres, C. A., Amano, T., Llactayo, W., & Leader-Williams, N. (2017). Conservation performance of different conservation governance regimes in the Peruvian Amazon. *Scientific Reports, 7*, 1–10.
Sundberg, J. 2015. Ethics, entanglement, and political ecology. The Routledge handbook of political ecology:117–126.
Tanhua, T., S. Pouliquen, J. Hausman, K. O'brien, P. Bricher, T. De Bruin, J. J. Buck, E. F. Burger, T. Carval, and K. S. Casey. 2019. Ocean FAIR data services. Frontiers in Marine Science 6:440.
de Vos, A., & Schwartz, M. W. (2022). Confronting parachute science in conservation. *Conservation Science and Practice, 4*.
Wade, R. 1994. Village republics. ICS Press.

Graham Epstein is a Research Assistant Professor at the University of Waterloo's School of Environment, Resources and Sustainability. His research explores the human dimensions of conservation and natural resource governance, focusing on collective action, compliance, and institutional performance. He collaborates with NGOs and government partners to inform policy design.

Rafaella Lobo is a PhD candidate at the Nicholas School of the Environment, Duke University. She is a critical social scientist whose research investigates the politics of international biodiversity governance, conservation policy, and the inclusion of Global South perspectives in global environmental negotiations. Her work combines qualitative political-economic analysis with active engagement in conservation and policy practice.

Evan J. Andrews is a Banting Postdoctoral Fellow in Geography at Memorial University of Newfoundland. His research centers on anticipatory governance and sustainability in small-scale fisheries, particularly in Newfoundland and Labrador. He employs transdisciplinary methods to engage stakeholders in shaping equitable ocean governance and is a coordinator for the Canadian hub of the "Too Big To Ignore" global fisheries research network.

Maria Battaglia is a PhD candidate at the University of Waterloo's School of Environment, Resources and Sustainability. Her research integrates economics and social science to explore sustainability under future climate scenarios, with a focus on the socioeconomic dynamics that foster sustainable practices. She is affiliated with the Environmental Change & Governance Group (ECGG).

Isabel Jorgensen is a PhD candidate whose research explores the ecological and social dimensions of desert lakes in the American West, including issues such as migratory bird declines and inequitable economic arrangements. She is interested in how agricultural transitions and water use conflict/cooperation play out under stressors like drought, especially in contexts where downstream / out-of-region demands compete with local environmental and social wellbeing. She also has been recognized via the P.E.O. Scholar Award for work that seeks to combine strong academic output with outreach or societal impact.

Ella-Kari Muhl is a marine conservation scientist and political ecologist affiliated with the School of Environment, Enterprise and Development at the University of Waterloo. Her research addresses equity, knowledge co-production, and governance of marine resources and protected areas. She emphasizes participatory methods and transdisciplinary approaches to advancing coastal community wellbeing and conservation outcomes.

Helena Shilomboleni is an assistant professor in the School of Environment, Resources and Sustainability at the University of Waterloo. Her research focuses on food security, sustainable agriculture, and climate adaptation in African contexts. She examines how governance arrangements and innovation systems shape pathways toward equitable and resilient food systems.

Open Access This chapter is licensed under the terms of the Creative Commons Attribution-NonCommercial-NoDerivatives 4.0 International License (http://creativecommons.org/licenses/by-nc-nd/4.0/), which permits any noncommercial use, sharing, distribution and reproduction in any medium or format, as long as you give appropriate credit to the original author(s) and the source, provide a link to the Creative Commons license and indicate if you modified the licensed material. You do not have permission under this license to share adapted material derived from this chapter or parts of it.

The images or other third party material in this chapter are included in the chapter's Creative Commons license, unless indicated otherwise in a credit line to the material. If material is not included in the chapter's Creative Commons license and your intended use is not permitted by statutory regulation or exceeds the permitted use, you will need to obtain permission directly from the copyright holder.

Ethical Dimensions and Practical Guidance for Transdisciplinary Research on the Commons

Jeremy Pittman, Gabriela Alonso Yáñez, Liana O. Anderson, Mauricio Castrejón, Guilherme Marques, Lily House Peters, and Micaela Trimble

Abstract The ethical dimension of transdisciplinary research is an emerging field of study with great importance for how we conduct research to address societal problems related to the commons. Transdisciplinary research typically aims to engage multiple knowledge systems from both academic and non-academic actors to contribute to science, policy, and practice. In some ways, transdisciplinary research represents an attempt to improve some of the ethical dimensions of science by valuing and integrating diverse knowledge systems. Nonetheless, the practice of transdisciplinary research opens new ethical issues during the three phases of transdisciplinary

J. Pittman (✉)
School of Planning, University of Waterloo, Waterloo, Canada
e-mail: jpittman@uwaterloo.ca

G. A. Yáñez
Werklund School of Education, Specialization, Curriculum and Learning, University of Calgary, Calgary, Canada
e-mail: galonsoy@ucalgary.ca

L. O. Anderson
Instituto Nacional de Pesquisas Espaciais, São José, Brazil
e-mail: liana.anderson@inpe.br

M. Castrejón
Universidad de las Américas-Programa Galápagos, Quito, Ecuador
e-mail: hugo.castrejon@udla.edu.ec

G. Marques
Institute of Hydraulic Research, Porto Alegre, Brazil
e-mail: guilherme.marques@ufrgs.br

L. H. Peters
Geography, California State University, California, USA
e-mail: lily.housepeters@csulb.edu

M. Trimble
Universidad de la República & South American Institute for Resilience and Sustainability Studies, Montevideo, Uruguay
e-mail: micaela.trimble@saras-institute.org

© The Author(s) 2026

S. Partelow (ed.), *Ethics and the Commons*, Schriften zu Leben und Gesellschaft - Series on Life and Society 3, https://doi.org/10.1007/978-3-032-24246-4_9

research—the (1) pre-proposal, (2) main research, and (3) feedback and dissemination stages, in addition to evaluation, which generally occurs across all stages of a project. During the pre-proposal phase, ethical issues relate to the identification of focal problems and the strategic, but not overburdening, engagement of local actors in the co-production of proposals. During the main research phase, ethical issues relate to balancing the needs of rigorous data collection and methodological approaches with the time, resources, and knowledge of local people. During the feedback and dissemination phase, ethical issues relate to who bears the burden of ensuring impact, how research deliverables can align with community members' or other partners' needs, and the blurry roles of researchers in engaging beyond the academy. We explore these issues in five case studies of transdisciplinary research projects in the Americas and detail the features of these ethical issues to provide guidance on how to address them. Our chapter contributes to the ongoing conversation regarding ethics in transdisciplinary research.

1 Introduction: Transdisciplinary Research as an Attempt to Improve the Ethics of Research

Transdisciplinary research aims to tackle problems of real-world importance through collaboration across disciplines and between diverse actors, such as researchers, practitioners, resource users, community members, and policy- or decision-makers (Augenstein et al., 2024; Lang et al., 2012). Transdisciplinary research is both a process of commoning (i.e., generating knowledge commons through collaboration) and it often aims to address problems related to the commons (e.g., biodiversity loss, climate change). Ethics is central to transdisciplinary research (Belcher et al., 2016; Reed et al., 2023). Ethical reflexivity is key to knowledge production, which alludes to the indivisibility of the epistemological and, ultimately, practical aspects of research (von Unger, 2021). A contemporary strand of transdisciplinary scholarship describes transdisciplinary approaches as grounded in foundational, long-standing literature with deep feminist, antiracist, decolonial, and critical social justice roots and commitments (Boogaard & Birgit, 2021; Rigolot, 2020; Staffa et al., 2022). In this view, knowledge production is a political commitment to advance social justice and equity and diversity in research practice. Researchers should apply methods and approaches in transdisciplinary research that enhance their capacities to critically examine situations and identify how they can avoid reproducing processes of inequality in all its dimensions (Alonso-Yanez et al., 2019; Boogaard & Birgit, 2021; Partelow et al., 2023; Porto de Albuquerque et al., 2023; von Unger, 2021).

Transdisciplinary research emerged in part due to ethical issues with previous types of research projects (Rigolot, 2020). In prior research, the predominance of Western science from the Global North and exclusion of alternative knowledge systems had become a major issue (Roux et al., 2017). This issue stemmed from the exclusion of diverse knowledge systems and types within global society, and these patterns of exclusion and power would often be reproduced in research projects (Goodchild, 2021; Zavala, 2013). Transdisciplinary research attempts to make the inclusion of diverse knowledge systems a focal task. The process for melding or weaving together different knowledge systems and expectations of the project is far from straightforward and depends on local cultural contexts, but nonetheless transdisciplinary researchers by definition are attempting to work across disciplines with local communities and to reduce the negative impacts of patterns of exclusion and power that exist in a society, including within academia (Rigolot, 2020; Scholz & Steiner, 2015). As such, in some ways, transdisciplinary research strives to be an ethical endeavor.

Nonetheless, the process of doing transdisciplinary research also raises ethical issues (Cockburn & Cundill, 2018). For instance, researchers attempting transdisciplinarity face several challenges related to how projects engage non-academics and community members, how projects value non-academic knowledge and incorporate it into research plans and timelines, and how the research process translates into mutually agreed upon and beneficial outcomes. Casal–Ribeiro et al. (2024) demonstrate how research fatigue of community members is a key ethical issue to tackle within transdisciplinary research projects. Transdisciplinary researchers also typically find themselves in non-traditional and often ambiguous roles, such as policy engagement and coordination (Gabriela Arnold, 2021). Often, the structures of academia challenge the transdisciplinary intent of researchers due to time constraints, disciplinary norms, and the ongoing legacies of colonial approaches to research (Manuel-Navarrete et al., 2021). The structures of society place similar burdens on research teams, who struggle to confront dominant power regimes and promote transformations towards sustainability (Saunders et al., 2016).

Ethics in the research process is not a new area of study, and ethical issues are inherent in transdisciplinarity (Strand, 2024). There have been countless researchers that have studied ethical dimensions of research from various perspectives. One that is particularly important for dealing with transdisciplinary research is the ethical space described by Willie Ermine, an Indigenous scholar (Ermine, 2007). Ermine is a leader in the field of research ethics, especially when it comes to working with Indigenous communities, but his lessons are relevant for working with any community. Ermine's ethical space sets the tone for developing ethical relationships in transdisciplinary research projects, and it is used in this chapter as a lens to view the transdisciplinary research processes in the respective projects (Ermine, 2007). Ermine's (2007) concept of the ethical space guides our analysis of issues and lessons from the research projects.

Ermine (2007,195) defines ethics as "the capacity to know what harms or enhances the well-being of sentient creatures". We believe this definition of ethics is important for guiding transdisciplinary research because essentially the main goal is to know and understand what can enhance wellbeing in our study sites, but we must be very careful and cautious to avoid harm on both the spaces and communities that are part of the study and the scientists themselves. Strand (2024) takes a similar stance and notes the potential for harm and extraction of knowledge as key ethical issues facing transdisciplinary research. Ermine's (2007) ethical space emerges when different groups or societies interact and must navigate various ethical issues related to avoiding harm and enhancing the wellbeing of one another.

Our goal with this chapter is to illuminate some of the ethical issues that are involved with transdisciplinary research as researchers and other participants traverse the ethical space that is inherent to and created by transdisciplinary research projects. Our analysis is based on experience conducting transdisciplinary research projects. We argue that, while transdisciplinary research in general is important to confront ethical issues in research, the process of conducting transdisciplinary research also presents numerous ethical issues for researchers. We use illustrative examples from recent transdisciplinary research projects throughout the chapter to demonstrate these issues. These lessons are aimed at other researchers and funding organizations.

2 Lessons from Five Transdisciplinary Research Projects in the Americas

We are drawing on lessons and experiences from five transdisciplinary research projects in the Americas, which were funded by the Inter-American Institute for Global Change Research's (IAI) Co-funded Small-Grants Program on the role of ecosystem services in adaptation to global change for human wellbeing (SGP-HW). These projects were conducted from 2019 to 2022 and cover a diverse range of topics and contexts related to the commons (Table 1).

First, the Multi-Actor Adaptation Plan to cope with Forests under Increasing Risk of Extensive Fires (**MAP-FIRE**) project aimed to investigate and develop tools to tackle increased fire threats to Amazonian ecosystems, its inhabitants, and the international commitments of Amazonian countries. This project developed collaborations between scientists and stakeholders, which produced a range of outputs including scientific papers to educational materials and online monitoring platforms to strengthen the knowledge flow and empowerment of social actors.

Second, the Transforming Water Governance in South America: From Reaction to Adaptation and Anticipation (**GovernAgua**) aimed to contribute to ecosystem service provisioning and human wellbeing associated with water, through a transdisciplinary research approach that sought to improve adaptive and anticipatory governance in South America. The project aimed to strengthen adaptive capacities, anticipation, co-creation of knowledge and social learning. The project worked in six watersheds: two in Argentina, two in Brazil and two in Uruguay, but a transdisciplinary approach was adopted mainly in three of them (due to time and resources constraints), where researchers worked collaboratively with members of basin committees - advisory boards for watershed management (Jacobi et al., 2022; Trimble et al., 2022).

Third, the Transdisciplinary Research Lab for Global Sustainability (**TRL-GS**) project is a multi-national transdisciplinary collaboration investigating how established teams negotiate the process of conducting place-based social-ecological

Table 1 Overview of focal projects

Project	Brief description	Commons	
		In research process	As the focus of research
MAP-FIRE: Multi-actor adaptation plan to cope with forests under increasing risk of extensive fires (SGP-HW 016)	Adaptation to wildfire risk in Bolivia, Brazil, and Peru.	Social and natural scientists from federal and state level, NGOs, state and municipal servers and school communities of three countries.	Strategies and tools to prevent fire risks and impacts on forest commons.
GovernAgua: Transforming water governance in South America: From reaction to adaptation and anticipation (SGP-HW 056)	Adaptation in water governance systems in Argentina, Brazil, and Uruguay.	Social, natural and interdisciplinary scientists, government organizations, social organizations (cooperatives, NGOs, farmers).	Water commons and crises of multiple types, and related governance (e.g. multistakeholder basin committees).
TRL-GS: Incorporating local and traditional knowledge systems: New insights for ecosystem services and transdisciplinary collaborations (SGP-HW 072)	Engaging diverse knowledge and value systems in transdisciplinary projects related to ecosystem services in Chile, Colombia, and Uruguay.	Social and natural scientists, education specialists, territorial defenders, local community members and activists, civil society, private property owners.	Implementation challenges at the local scale related to conservation of biodiversity and natural commons
HYDRO: Improving the governance of the floodplain in over-built river basins (SGP-HW 091)	Floodplain governance in Brazil, Canada, and Chile.	Social and natural scientists, energy providers, government decision makers	Water and energy commons and decision making.
Fish-MES: Small-scale fisheries and marine ecosystem services: Adaptation and transformation to secure human wellbeing (SGP-HW 017)	Adaptation to environmental change in small-scale fisheries in Argentina, Brazil, Ecuador, and Uruguay.	Ecologists, oceanographers, social scientists, fishers, fishers' organizations, government agencies.	Coastal-marine commons facing climate change and biodiversity loss.

research. The team is composed of citizens of Latin American countries, which offers a unique perspective for regional focus. The team's program of research concerns action-oriented work focused on understanding the factors and conditions that influence how teams co-produce integrated, action-oriented social-ecological knowledge in networks that include knowledge keepers, local community members and academics.

Fourth, the Adaptation Pathways to Reconcile Hydropower Generation and Ecosystem Services project (**HYDRO**) was motivated by the growing challenge to protect remaining water ecosystems in the context of energy hungry societies relying on different power systems. Hydropower is a key element in many of these systems, given its clear advantage over thermoelectric power sources in terms of GHG emissions and their impact on climate change (Rosa et al., 2016). However, in over-built basins, hydropower reservoirs modify the natural flow regime and disrupt the natural equilibrium of aquatic ecosystems. Although technical solutions for designing flow regimes are promising, they still lack better understanding of tradeoffs and implementation within the governance framework. The project addressed this gap by developing methods to support the identification of improved reservoir operating policies, assess energy tradeoffs under climate change and to equitably distribute the tradeoffs among society. The main conclusions provide general and specific recommendations for water policies. While these are expected to contribute to a broader impact to society, they are also fraught with ethical issues given their strong transdisciplinary nature.

Fifth, the Small-scale Fisheries and Marine Ecosystem Services: Adaptation and Transformation to Secure Human Wellbeing project focused on community-based fisheries in Uruguay, Brazil, Argentina, and Ecuador (**Fish-MES**). This project conducted a comprehensive examination of the effects of climate change on small-scale fisheries and the ecosystem services vital to coastal communities in regions undergoing rapid warming. It leveraged an international collaboration of researchers and practitioners across various disciplines and study sites to confront the challenges of fostering climate-resilient and adaptable small-scale fisheries. By synthesizing data from different locales, the study recommended management strategies and adaptation actions, culminating in customized guidance for the sustainable use and conservation of sensitive ecosystems.

3 The Ethical Issues Emerging in Transdisciplinary Research

We distill ethical issues and lessons for the different phases of transdisciplinary research. According to (Cockburn & Cundill, 2018; Lang et al., 2012), the transdisciplinary research process involves three main phases (Fig. 1). First, there is the pre-proposal phase. In this phase, the researchers are developing proposals and partnerships to work together with non-academic actors to co-produce a set of research questions and objectives that can be used to develop a research proposal to attract funding. Second, there is the main research phase. The main research phase occurs after a proposal has been successful or a project can move forward, and during this phase is when most of the data collection, exchanges with the stakeholders and communities, analysis, and synthesis takes place. This is a phase of a research project where most of the practice of doing 'research' and building trust among the team members and with the stakeholders occurs. Third, there is the feedback and dissemination phase, which can co-occur with phase 2. In this phase of the research, the researchers and other partners are attempting to refine the findings from research projects, connect them with actionable strategies, and help steward the impact of the research project to policy and practice. We highlight that interaction, collaboration, assessment, and evaluation are ongoing processes that occur through the three phases. These phases and their ethical issues through the lens of the ethical space are summarized in Table 2.

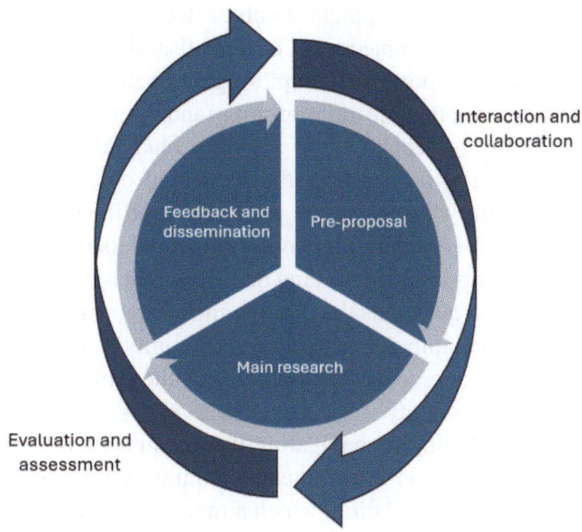

Fig. 1 The transdisciplinary research process (adapted from Cockburn & Cundill, 2018). The image depicts a conceptual diagram, a circle divided into three sections (feedback and dissemination, pre-proposal, main research) with circulating arrows around it that are labelled interaction and collaboration, and evaluation and assessment

Table 2 The main ethical questions across research phases

Pre-Proposal	Main research	Feedback and dissemination
● Who to include? Whose knowledge leads? ● What problems of the commons will be tackled? How? ● Whose time and resources? Who is expected to benefit? How will resources be distributed? ● What does the funding or call allow to be included?	● How to dismantle problematic power relations? ● How to respect diverse knowledge? How to incorporate ongoing advice and address emerging concerns? ● How to meaningfully and efficiently communicate, engage, and interact? ● How to make progress on the research project and balance community members' and partners' time commitments?	● How to develop and deliver products of use to different actors (e.g. society, policy makers)? ● What roles are appropriate for the researcher given community priorities, capacities, and perspectives? ● Is the timing of the project/funding aligned with the time needed to develop the products?

3.1 Pre-Proposal Phase

The development of a proposal is usually the first step in many research projects, which typically involves choosing study areas, understanding the contexts of these areas, and bringing together the right group of people to develop a project in these areas (Horcea-Milcu et al., 2022). Addressing ethical issues at this stage is important to ensure the project is capable of positively affecting the livelihoods of local communities. The risk of alienating local stakeholders through top-down processes will likely limit the project's positive impacts to society.

Proposal development involves determining the focal research questions of the project and the objectives. In transdisciplinary research, the pre-proposal phase is incredibly important for exploring and balancing power relations within the project and developing the relationships and protocols that will lead to successful and impactful projects. One of the most challenging power issues to deal with during this phase is determining the focus and objectives of the project, which involves navigating whose ideas or priorities contribute. This stage involves balancing the needs of communities and other stakeholders with the priorities of funding agencies and the important or enticing questions that remain unresolved in academic literature (i.e., those questions that are the most interesting from an academic perspective). It is an ethical imperative that the community's and stakeholders' needs are at the forefront in this process, but often researchers must strike a balance between focusing on the community's or stakeholders' needs, meeting the objectives and priorities of a funding call, and navigating the agendas of various types of organizations (e.g., government, non-government, community-based). This can often involve various processes of translation, which aim to find synergies between the interests and needs of various stakeholders and rights-holders.

This phase requires a lot of coordination, meetings, and other forms of interaction, and it is important to account for the time necessary to complete these processes well. However, projects typically do not have funding at this phase. While it is important to involve non-academic actors from the stage of drafting the proposal until the stage of disseminating and implementing the results, in many instances, there are limited resources to do so. The expansion of online and remote options for interaction helps during this phase, but many marginalized people, either geographically or otherwise, may not have access to powerful enough internet to support these interactions. Additionally, even within marginalized communities, there are also issues related to unbalanced power, funds and access. Essentially, site visits are often the only means to help ensure project development includes all voices to avoid reproducing existing disparities. The IAI SGP-HW program provided funding for workshops during this phase, which is quite uncommon, but incredibly beneficial based on our experiences.

Another important ethical dilemma arises during this phase regarding the need to balance community participation with the demands of community members' time spent on co-developing or co-producing a transdisciplinary research proposal. The proposal development phase in transdisciplinary research is slightly different from other forms of research, since the engagement of community members and their knowledge is an integral part of developing the proposal. However, community members' time is often at a premium and proposal development can be an incredibly time-consuming process. Researchers working with communities to develop proposals must be mindful of and respect the multiple demands that are often placed on community members' time. In this regard, researchers and community members must weigh the potential returns on investment for communities and the likelihood that proposals and deliverables will be successful and useful for the locals. Researchers should make all this transparent to community members and then the community members can make their own informed decisions about participating in the proposal development phase or not.

A final and related ethical problem refers to the need for matching funding in the form of cash or in-kind contributions, which is a requirement of many funding agencies. While the provision of in-kind or cash can be positive for the project and it can improve the buy-in or ownership that non-academic groups have over the project, it can also be somewhat extractive and usually places added demands for resources on communities and other groups and organizations. Furthermore, it can be a hindrance to developing meaningful relationships with community members and groups, especially if researchers are developing collaborations with new communities or in new places.

3.2 Main Research Phase

Much of the ethical issues during the main research phase relate to initiatives aimed at strengthening academic spaces by prioritizing principles of equity, diversity, inclusion, decolonization, and justice (Eaton, 2022; Trisos et al., 2021). The TRL-GS team, composed mostly of Latin American citizens, welcomed these commitments, which became central to their ethical engagement during their research project. Serious engagement with principles of equity, diversity, inclusion, decolonization, and justice is not static or merely instituted once and then forgotten about; rather, serious engagement represents an allegiance that must be periodically refreshed among participants considering new experiences and learnings. As a team, the TRL-GS project offers two concrete examples of how they operationalized efforts to cultivate equitable relations, offer authentic spaces for diverse perspectives and views, and respectfully include diverse ways of knowing and being in the world relevant to our contexts.

First, research processes and the procedures themselves also must be ethically adapted to align with local contexts and acknowledge the positionality of the researchers. An example from the TRL-GS project highlights this need in the context of working with Indigenous peoples. Inspired by the 94 recommended calls to action in the Truth and Reconciliation Commission of Canada (Truth & Reconciliation Commission of Canada 2015), land acknowledgements are honest attempts to recognize the traditional First Nations, Métis, and Inuit territories of a place. Because of a careful reflection by the TRL-GS team, with active immigrant experiences in Canada, they supplemented their land acknowledgement with a land commitment that states their roles as guests in the land. They explicitly acknowledge their continuous dialogue with local knowledge holders, with whom they have established relationships, facilitating conversations about their moral and ethical standpoint in research and activism.

Second, research participants and partners need access to opportunities within the project. A central part of the TRL-GS project is designing learning and capacity building initiatives for professionals involved in global change research with diverse epistemological affiliations and sources of knowledge. A key step in designing authentic transdisciplinary projects requires an intentional work of 'decentering academia", which the TRL-GS project defined as a horizontal process through which research decisions are made, in dialogue with research partners from other sectors (e.g., non-governmental organizations, policymakers, community leaders, and so on) (Alonso-Yanez et al., 2019). This tactic limits the potential for a single body of knowledge to powerfully direct decisions about the research process and associated allocation of resources (Alonso-Yanez et al., 2019). It involves facilitating conversations that critique how universities and academia often establish implicit normative and evaluative criteria regarding knowledge.

The generation of data on transdisciplinary projects also contributes to ethical challenges. Transdisciplinary projects offer a unique approach to generating findings essential for producing knowledge through unconventional methods. By actively involving local stakeholders throughout the research process, these projects are better positioned to inform policies that effectively meet the needs of communities. However, ethical considerations must be at the forefront of these efforts, particularly regarding data access and usage (M. Mastrangelo, personal communication March 28, 2024). Questions arise about who will access the generated information and for what purposes. These considerations are critical to the integrity of knowledge generation and highlight significant ethical issues. It is vital to incorporate a data-sharing plan into the early stages of transdisciplinary (TD) project planning to address many of these ethical challenges. Rather than viewing data sharing as a final requirement, integrating it from the outset can prevent potential ethical issues related to privacy, data sensitivity, and overall data usability. By designing data-sharing strategies from the beginning, researchers can comply with ethical standards and enhance the relevance and applicability of their findings (M. Mastrangelo, personal communication March 28, 2024). This proactive approach would help mitigate complications related to ethical conduct in research, making the entire process more effective while ensuring that the knowledge generated is responsible and transparently shared. Ultimately, prioritizing these ethical dimensions fosters trust among stakeholders and leads to more sustainable outcomes for the communities involved.

Since communication is a key ingredient for the success of a project and the management of ethical concerns, a transdisciplinary team must adopt multiple means of communication, since different actors may prefer different means. Additionally, projects should design participatory workshops and other interactive forums in consultation with the participants themselves, since this can enhance their engagement and affect ethical outcomes. However, flexibility in communication processes and research activities is a must, which—for the focal projects—was underscored by the COVID-19 pandemic (Hall et al., 2021; Kulundu-Bolus et al. 2021).

For the focal projects, the pandemic not only affected peoples' lives but also on the plans of activities of the projects, which needed to be changed dramatically. In the worst times of the pandemic, patience and flexibility were key because the projects were not necessarily a priority for some participants and partners, who were going through very hard times, with implications for the funders and research teams. The projects learned about the potential of virtual tools for communication among team members and with stakeholders. They also found out the need for skills in virtual facilitation of participatory workshops, and they learned to use various tools, including radio. Some activities that we had planned face-to-face became virtual, others could not be replaced, and new virtual activities or materials were developed to overcome these limitations (e.g., https://efogo.com.br/). These efforts were important to adhere to the ethical principles around participation inherent in a transdisciplinary project (Hall et al., 2021; Kulundu-Bolus et al. 2021).

The IAI's flexibility on how to invest the project grant increased the projects' capacity to deal with pandemic related ethical issues. In many cases, interactions with remote communities were difficult to complete either due to the communities' geographical situation or the associated costs. For the MAP-Fire project, the initial plan was to perform a series of three workshops with school communities to co-create and develop educational activities related to fire risks and impacts in the Amazon (de Paula et al., 2021a, 2021b). However, with the advent of the COVID-19 pandemic, there were readjustments in the structure of the school activities and a two-pronged approach emerged. For the first prong of the approach, virtual meetings replaced the face-to-face workshops that were initially planned, which helped keep progress going, but unfortunately, people with no or limited internet access could not join. For the second prong of the approach, the researchers created physical activity guidelines and books, which could be sent to the schools to foster engagement. These materials were in line with adapted scientific methodologies, aiming to awaken students to "do science, investigate and report results". This material was available in both Portuguese and Spanish to broaden accessibility. Switching directions in this way required flexibility in the project budgets.

An additional lesson learned from multiple projects considering the COVID-19 pandemic was the inherent challenges of orchestrating transdisciplinary research at a multinational level without the direct collection of data (Hall et al., 2021). The endeavor of integrating knowledge from existing datasets from various contexts, each with its unique set of circumstances, methodologies, and outcomes, underscored the difficulties of conducting research across diverse geographical and cultural landscapes during an unprecedented global health crisis. This approach, while resourceful and innovative, introduced significant logistical, financial, and communication challenges, which were amplified by the disruptions caused by the pandemic. Logistically, the projects faced the task of aligning and synthesizing data from different countries, each with its regulatory, environmental, and social nuances, in a period marked by travel restrictions, lockdowns, and varying degrees of access to research facilities.

The pandemic added a layer of complexity to the research, and ultimately impacted how ethical issues were navigated (Kulundu-Bolus et al. 2021).

The COVID-19 pandemic significantly impacted communication, presenting unique challenges in fostering collaboration and knowledge exchange among team members isolated by global travel restrictions and local lockdown measures. Establishing a shared understanding and cohesive approach to integrating diverse datasets necessitated an even greater reliance on digital communication tools. This effort was further complicated by the isolation and digital fatigue experienced by many researchers, making clear and efficient communication more critical than ever. Despite these additional challenges imposed by the pandemic, the projects highlight the intricate balance required to manage transdisciplinary research across varied international contexts during a global health crisis. It demonstrated the importance of flexible analytical approaches to overcome the hurdles of remote collaboration. This approach also emphasized the potential of using existing datasets to address global challenges, provided there is a strong framework for collaboration that can accommodate the complexities of working across disciplines, languages, national boundaries, and unprecedented global disruptions.

3.3 Feedback and Dissemination Phase

The third phase of transdisciplinary research projects involves gathering feedback from community members and disseminating the research findings and or co-producing products or outputs useful for the locals and non-academic stakeholders. During this phase, one of the most important ethical challenges is developing locally appropriate deliverables from the project, which can help achieve a broad range of goals. Researchers are often focused on publishing papers in academic journals, which have broad benefits, but are not necessarily the types of deliverables that community members or other partners need to achieve the outcomes or goals that they desire. Researchers often must produce deliverables that are valuable to the research partners, but not necessarily valued within the academic world. Nonetheless, researchers have the ethical responsibility during this phase to pay close attention to the needs of their research partners to ensure that a broad range of research products are developed. For example, in the MAP-Fire project, there were 1000 copies of a printed ebook and fact sheets developed to facilitate capacity building. These materials were crucial to the communities and resulted from many collaborations to provide the physical space, food supply, and transport of the Map-Fire team to enhance training activities with communities and organizations in different regions.

Beyond the development of diverse research and deliverables or products, this step also involves using these products to make a meaningful change in the world and researchers must navigate their ethical role and power in the process of having impact (Fuentes, 2024). In many cases, it is common for researchers to take on more of an advocacy or lobbying role in this phase. And in other cases, it is also common for researchers to leave this work up to their partners. Both are appropriate responses, but what is important is that the researchers allow the community members and other partners to guide how the researchers can best participate in this process. At times, advocacy by researchers is an expectation of the funder, but such advocacy may not be ethical, if it does not align with community members' and partners' views of the role of researchers. Researchers do not always have the in-depth knowledge and experience of how different management or governance systems work to engage effectively with these processes, and well-intentioned efforts can have undesirable effects if not handled properly. It is also important that the researchers' actions do not bring unnecessary risks to research participants and partners, especially if the researchers' efforts shine a spotlight on certain issues.

For the HYDRO project, when setting up the mathematical model for reservoir system optimization, the very discussion on which competing uses to include raises ethical issues and relies on knowledge from other fields, such as sociology and economics. The project sought to bring in a transdisciplinary team to make sure different perspectives were discussed. After results were produced, the negotiation process to define ecosystem recovery strategies to minimize trade-offs and improve chances of effective implementation also depended on well-balanced communication and capacity building at the local level. Recovery strategies will involve reservoir operation, which affects local communities and energy production both positively and negatively. Local and regional communities need to be engaged in a discussion that can be complex and far-reaching, which goes beyond the project. Local communities and stakeholders are a key part of the problem formulation, negotiation to select the best solution and even outcomes monitoring without creating unnecessary political risk.

In the TRL-GS project, academics who align with a social justice-oriented commitment engage in a process that invites far-reaching reflections about research practices of knowledge production, ethical reflexivity, and research aims. This involves problematizing power relationships between researchers and the "researched" and tackling questions such as "whose story is told, how, by whom, and for whose benefit?" (Clarke, Friese & Washburn, 2017).

An additional ethical challenge emerges in this phase, which relates to researchers navigating diverse roles within the governance systems in which they participate (Fuentes, 2024; Manuel-Navarrete et al., 2021; Wittmayer & Schäpke, 2014). It is common during transdisciplinary research to develop strong connections to various partners and organizations, and at times researchers even become directly engaged with these different organizations as board members or in other roles. Invitations to these roles likely reflect the trust that partners have with the researcher, and it allows the researcher to develop better projects that will more likely be impactful. However, it also places the researcher very close to the system that they are studying, which usually means that researchers are participating in the various political dimensions of their study sites. To some, this is the ultimate goal: to interact within the political sphere and achieve impact. To others, this brings into question the objectivity of the researcher and their research, which can have implications for the legitimacy of project findings. There is no easy answer, and each situation is likely different.

4 Summary and Path Forward

There are three main components of an ethical pathway for transdisciplinary research, which we believe our experiences across the five projects highlight. First, research projects require capacity and investment in the pre-proposal phase to effectively engage with communities and other partners and establish ethical procedures within project governance. This phase of research is often under-funded and overlooked, despite being crucial, and enhanced focus on this phase can lead to improved ethical outcomes in later stages of projects (Horcea-Milcu et al., 2022; Wolff et al., 2019). Ethics are the responsibilities of both funders and researchers. Funders determine what is included in the call for proposals and what can be funded. Researchers are responsible for ensuring the appropriate inclusion of non-academic actors in the three phases.

Second, flexibility to accommodate community members' and partners' changing needs and priorities is crucial during the main research phase. In some cases, this flexibility could go beyond simple plans for research activities and involve complete repurposing of existing project resources for new and emerging issues that communities must face. Ethics may lead to prioritizing the communities' and partners' needs and timelines, which are not always aligned with formal procedures (e.g., permits, deadlines).

Third, research projects require sustained commitment during the development of deliverables, feedback and dissemination phase to realize fully the potential benefits of the projects' results to science and local needs. This often involves researchers working on traditionally non-academic outputs and deliverables, or participating in traditionally non-academic forums, but nonetheless, it is essential from an ethical perspective for researchers to develop these deliverables and participate in these forums even though they may not be valued within the academy. Ethics is central to the practice of transdisciplinary research, and we hope to underscore the importance of ethical considerations for the successful implementation of transdisciplinary projects.

Acknowledgements The authors would like to thank the Inter-American Research Institute for Global Change Research (IAI) for funding support through grants SGP-HW 016 (Anderson), SGP-HW 056 (Trimble), SGP-HW 017 (Pittman and Castrejón), SGP-HW 091 (Marques), and SGP-HW 072 (Alonso Yáñez and House Peters). Anderson also acknowledges funding support from the Sao Paulo Research Foundation (FAPESP) grant number 2021/07660-2.

References

Alonso-Yanez, G., House-Peters, L., Garcia-Cartagena, M., Bonelli, S., Lorenzo-Arana, I., & Ohira, M. (2019). Mobilizing transdisciplinary collaborations: Collective reflections on decentering academia in knowledge production. *Global Sustainability, 2*, e5. https://doi.org/10.1017/sus.2019.2

Augenstein, K., Lam, D. P., Horcea-Milcu, A.-I., Bernert, P., Charli-Joseph, L., Cockburn, J., Kampfmann, T., Pereira, L. M., & Sellberg, M. M. (2024). Five priorities to advance transformative transdisciplinary research. *Current Opinion in Environmental Sustainability, 68*, 101438. https://doi.org/10.1016/j.cosust.2024.101438

Belcher, B. M., Rasmussen, K. E., Kemshaw, M. R., & Zornes, D. A. (2016). Defining and assessing research quality in a transdisciplinary context. *Research Evaluation, 25*(1), 1–17. https://doi.org/10.1093/reseval/rvv025

Boogaard, D. L., & Birgit, K. (2021). Making transdisciplinarity work: An epistemology of inclusive development and innovation. In *The politics of knowledge in inclusive development and innovation*. Routledge.

Casal–Ribeiro, M., Peixoto, U. I., Medeiros–Leal, W., & Santos, R. (2024). Research fatigue's impact on small-scale fishers' engagement: A case-study from Azores fisheries. *Marine Policy, 170*, 106404. https://doi.org/10.1016/j.marpol.2024.106404

Cockburn, J., & Cundill, G. (2018). Ethics in Transdisciplinary Research: Reflections on the Implications of 'Science with Society.' In C. I. Macleod, J. Marx, P. Mnyaka, & G. J. Treharne (Eds.), The Palgrave handbook of ethics in critical research (pp. 81–97). Springer International Publishing. https://doi.org/10.1007/978-3-319-74721-7_6.

Eaton, S. E. (2022). New priorities for academic integrity: Equity, diversity, inclusion, decolonization and indigenization. *International Journal for Educational Integrity, 18(1), Article 1*. https://doi.org/10.1007/s40979-022-00105-0

Ermine, W. (2007). The ethical space of engagement. *Indigenous Law Journal, 6*, 193.

Fuentes, A. (2024). Scientists as political advocates. *Science, 386*(6724), eadt 7194. https://doi.org/10.1126/science.adt7194

Gabriela Arnold, M. (2021). The challenging role of researchers coping with tensions, dilemmas and paradoxes in transdisciplinary settings. *Sustainable Development.*. https://onlinelibrary-wiley-com.proxy.lib.uwaterloo.ca/doi/full/10.1002/sd.2277

Goodchild, M. (2021). Relational systems thinking: That's how change is going to come, from our earth mother. *Journal of awareness-based systems. Change, 1(1), Article 1*. https://doi.org/10.47061/jabsc.v1i1.577

Hall, J., Gaved, M., & Sargent, J. (2021). Participatory research approaches in times of Covid-19: A narrative literature review. *International Journal of Qualitative Methods, 20*, 16094069211010087. https://doi.org/10.1177/16094069211010087

Horcea-Milcu, A.-I., Leventon, J., & Lang, D. J. (2022). Making transdisciplinarity happen: Phase 0, or before the beginning. *Environmental Science & Policy, 136*, 187–197. https://doi.org/10.1016/j.envsci.2022.05.019

Interviewee M. M. Mastrangelo, personal communication, March 28, 2024.

Jacobi, P. R., Trimble, M., & Dias Tadeu, N. (2022). Dimensiones inter y transdisciplinarias del Proyecto GovernAgua: aprendizajes y desafíos. *Diálogos Socioambientais, 5*(13), 37–43. https://periodicos.ufabc.edu.br/index.php/dialogossocioambientais/article/view/687

Lang, D. J., Wiek, A., Bergmann, M., Stauffacher, M., Martens, P., Moll, P., Swilling, M., & Thomas, C. J. (2012). Transdisciplinary research in sustainability science: Practice, principles, and challenges. *Sustainability Science, 7*(SUPPL. 1), 25–43. https://doi.org/10.1007/s11625-011-0149-x

Manuel-Navarrete, D., Buzinde, C., & Swanson, T. (2021). Fostering horizontal knowledge co-production with indigenous people by leveraging researchers' transdisciplinary intentions. *Ecology and Society, 26*(2). https://doi.org/10.5751/ES-12265-260222

Partelow, S., Schlüter, A., Ban, N. C., Batterbury, S., Bavinck, M., Bennett, N. J., Bleischwitz, R., Blythe, J., Bogusz, T., Breckwoldt, A., Cinner, J. E., Glaser, M., Govan, H., Gruby, R., Hatje, V., Hornidge, A.-K., Hovelsrud, G. K., Kittinger, J. N., Kluger, L. C., et al. (2023). Five social science intervention areas for ocean sustainability initiatives. *npj Ocean Sustainability, 2*(1), 1–11. https://doi.org/10.1038/s44183-023-00032-8

de Paula, Y.; Pismel, G.; dos Reis, J.B.C.; Anderson, L.O. 2021a "Guia de Atividades: É Fogo! Amazônia" 1ed. ISBN: 978-65-00-33309-1.

de Paula, Y.; Pismel, G.; dos Reis, J.B.C.; Anderson, L.O. 2021b "Guía de ACtividades: !Fuego! Amazonia" 1ed. ISBN:978–65–84510-12-8.

Porto de Albuquerque, J., Anderson, L., Calvillo, N., Cattino, M., Clarke, A., Cunha, M. A., Degrossi, L. C., Garde-Hansen, J., Klonner, C., Lima-Silva, F., Marchezini, V., Martins, M. H. da M., Pajarito Grajales, D., Pitidis, V., Rizwan, M., Tkacz, N., & Trajber, R. (2023). Dialogic data innovations for sustainability transformations and flood resilience: The case for waterproofing data. Global Environmental Change, 82, 102730. https://doi.org/10.1016/j.gloenvcha.2023.102730.

Reed, M. G., Robson, J. P., Campos Rivera, M., Chapela, F., Davidson-Hunt, I., Friedrichsen, P., Haine, E., Johnston, A. B. D., Lichtenstein, G., Lynes, L. S., Oloko, M., Sánchez Luja, M., Shackleton, S., Soriano, M., Sosa Peréz, F., & Vasseur, L. (2023). Guiding principles for transdisciplinary sustainability research and practice. *People and Nature, 5*(4), 1094–1109. https://doi.org/10.1002/pan3.10496

Rigolot, C. (2020). Transdisciplinarity as a discipline and a way of being: Complementarities and creative tensions. *Humanities and social sciences* Communications, *7*(1), Article 1. https://doi.org/10.1057/s41599-020-00598-5.

Rosa, L. P., Santos, M. A. dos, Gesteira, C., & Xavier, A. E. (2016). A model for the data extrapolation of greenhouse gas emissions in the Brazilian hydroelectric system. Environmental Research Letters, 11(6), 064012. https://doi.org/10.1088/1748-9326/11/6/064012.

Roux, D. J., Nel, J. L., Cundill, G., O'Farrell, P., & Fabricius, C. (2017). Transdisciplinary research for systemic change: Who to learn with, what to learn about and how to learn. *Sustainability Science, 12*(5), 711–726. https://doi.org/10.1007/s11625-017-0446-0

Saunders, F. P., Gallardo-Fernández, G. L., Van Tuyen, T., Raemaekers, S., Marciniak, B., & Plá, R. D. (2016). Transformation of small-scale fisheries—Critical transdisciplinary challenges and possibilities. *Current Opinion in Environmental Sustainability, 20*, 26–31. https://doi.org/10.1016/j.cosust.2016.04.005

Scholz, R. W., & Steiner, G. (2015). The real type and ideal type of transdisciplinary processes: Part II—What constraints and obstacles do we meet in practice? *Sustainability Science, 10*(4), 653–671. https://doi.org/10.1007/s11625-015-0327-3

Staffa, R. K., Riechers, M., & Martín-López, B. (2022). A feminist ethos for caring knowledge production in transdisciplinary sustainability science. *Sustainability Science, 17*(1), 45–63. https://doi.org/10.1007/s11625-021-01064-0

Strand, M. (2024). Advancing transdisciplinarity as an epistemology for more ethical engagements with fishing communities. *ICES Journal of Marine Science, fsae148.* https://doi.org/10.1093/icesjms/fsae148

Trimble, M., Olivier, T., Anjos, L. A. P., Tadeu, N. D., Giordano, G., Mac Donnell, L., Laura, R., Salvadores, F., Santana-Chaves, I., Torres, P. H. C., Pascual, M., Jacobi, P. R., Mazzeo, N., Zurbriggen, C., Garrido, L., Jobbágy, E., & Pahl-Wostl, C. (2022). How do basin committees deal with water crises? Reflections for adaptive water governance from South America. *Ecology and Society, 27*(2), 42. https://doi.org/10.5751/ES-13356-270242

Trisos, C. H., Auerbach, J., & Katti, M. (2021). Decoloniality and anti-oppressive practices for a more ethical ecology. *Nature Ecology & Evolution, 5*(9), 1205–1212. https://doi.org/10.1038/s41559-021-01460-w

von Unger, H. (2021). Ethical reflexivity as research practice. *Historical Social Research / Historische Sozialforschung, 46*(2), 186–204.

Wittmayer, J. M., & Schäpke, N. (2014). Action, research and participation: Roles of researchers in sustainability transitions. *Sustainability Science, 9*(4), 483–496. https://doi.org/10.1007/s11625-014-0258-4

Wolff, M., Cockburn, J., De Wet, C., Carlos Bezerra, J., Weaver, M., Finca, A., De Vos, A., Ralekhetla, M., Libala, N., Mkabile, Q., Odume, O., & Palmer, C. (2019). Exploring and expanding transdisciplinary research for sustainable and just natural resource management. *Ecology and Society, 24*(4). https://doi.org/10.5751/ES-11077-240414

Zavala, M. (2013). What do we mean by decolonizing research strategies? Lessons from decolonizing, Indigenous Research Projects in New Zealand and Latin America. *Education Faculty Articles and Research.* https://digitalcommons.chapman.edu/education_articles/106

Jeremy Pittman is an associate professor in the School of Planning, University of Waterloo. His work focuses on governance in social–ecological systems, with emphasis on land–sea interfaces, species-at-risk conservation, and cross-scale institutional coordination. He uses mixed methods and network approaches to study how governance arrangements affect ecological and social outcomes.

Gabriela Alonso-Yáñez is an associate professor in the Werklund School of Education at the University of Calgary. Her research explores education for sustainability, transdisciplinary learning, and the co-production of socio-ecological knowledge. She studies how educational processes and partnerships can enable integrated, action-oriented approaches to sustainability.

Liana Oighenstein Anderson is a senior researcher at the National Institute for Space Research (INPE) in Brazil, affiliated with the Laboratory of Aquatic Systems Instrumentation (LabISA). Her work focuses on land use and land cover change, fire ecology, and remote sensing applications in the Amazon region. She has contributed significantly to understanding fire-mediated carbon emissions and climate impacts in tropical ecosystems.

Mauricio Castrejón is a researcher at Universidad de las Américas in Ecuador, where he focuses on marine conservation and sustainable fisheries in the Galápagos Islands. His work emphasizes ecosystem-based management and the role of community-led governance in marine protected areas. He advocates for policy frameworks that balance ecological protection with artisanal fishing livelihoods.

Guilherme Marques is an associate professor at the Institute of Hydraulic Research, Universidade Federal do Rio Grande do Sul, Brazil. His research focuses on water resources systems analysis, reservoir and hydropower optimization, and the water–energy–food nexus. He integrates modelling and decision-support tools to inform integrated water management and planning under uncertainty and climate change.

Lily House-Peters is a human-environment geographer and political ecologist at the Department of Geography, California State University, Long Beach. Her research examines water governance, environmental justice, and climate vulnerability, using mixed qualitative and quantitative methods. She investigates how social, technological, and institutional factors shape environmental resilience and equity.

Micaela Trimble is a researcher at the Universidad de la República and the South American Institute for Resilience and Sustainability Studies (SARAS) in Uruguay. Her work centers on participatory and adaptive water governance, co-management of fisheries, and social–ecological resilience. She leads research and cooperation initiatives on stakeholder participation and sustainability transitions in South America.

Open Access This chapter is licensed under the terms of the Creative Commons Attribution-NonCommercial-NoDerivatives 4.0 International License (http://creativecommons.org/licenses/by-nc-nd/4.0/), which permits any noncommercial use, sharing, distribution and reproduction in any medium or format, as long as you give appropriate credit to the original author(s) and the source, provide a link to the Creative Commons license and indicate if you modified the licensed material. You do not have permission under this license to share adapted material derived from this chapter or parts of it.

The images or other third party material in this chapter are included in the chapter's Creative Commons license, unless indicated otherwise in a credit line to the material. If material is not included in the chapter's Creative Commons license and your intended use is not permitted by statutory regulation or exceeds the permitted use, you will need to obtain permission directly from the copyright holder.

Commons as Boundary Objects for Ethical Awareness: Key Questions and Directions Forward

Stefan Partelow, Guido Caniglia, Floriane Clement, Erin Rizzato Devlin, Graham Epstein, Svein Jentoft, Christopher Luederitz, Jeremy Pittman, and Christiane Woopen

Abstract As a catalyst for continued research on ethics and the commons, this chapter reflects on current ethical issues, key questions, and directions forward. We begin by arguing for the need to build ethical awareness, and then outline some of the ethical claims linked to the commons and the potential claims of different ethical theories about the commons. We believe there is a plurality of ethical perspectives and values worth considering across the different scholarly fields, practice communities, and contexts in which we observe and engage with them. We see many different areas where commons, common property, and commoning raise important ethical issues in our current day-to-day lives, neighborhoods, communities, and globally, and also within larger societal debates on economic, political, and environmental issues. These require continued debate and engagement in research, governance, and civil society, through which commons and commoning offer useful perspectives. This final chapter of the book aims to pinpoint these key areas of interest for continued engagement.

1 Are Commons Boundary Objects? Creating Spaces for Ethical Dialogue

We argue that commons offer spaces for ethical dialogue because they are boundary objects. This is because they almost always exist between different dichotomies or tensions in society (Table 1), and because of this, they bring together and relate to different ethical approaches and theories (Table 3). Sitting at the boundary can enable

S. Partelow (✉) · C. Woopen
Center for Life Ethics, University of Bonn, Bonn, Germany
e-mail: stefan.partelow@uni-bonn.de

C. Woopen
e-mail: chwoopen@uni-bonn.de

S. Partelow
Faculty of Agricultural, Nutritional and Engineering Sciences (AEI), University of Bonn, Bonn, Germany

© The Author(s) 2026

S. Partelow (ed.), *Ethics and the Commons*, Schriften zu Leben und Gesellschaft - Series on Life and Society 3, https://doi.org/10.1007/978-3-032-24246-4_10

dialogue and exchange, and recognition of pluralities. Commons bring people, ideas, resources, histories and possible futures together around shared communities, spaces, resources, or institutions of interest. Due to this, the ethics entangled in commons governance and commoning can be manifold and pluralistic, perhaps more than other governable entities or processes of governance. To develop the content in this chapter, we have conducted an iterative and reflective process among chapter authors of this book to distill the core issues and questions for future research and debate, based on our respective chapters, collective experiences, and knowledge of the literature. This began with the effort to write the chapters in line with the aim of the book, then in the form of a focused workshop, and continued as a collaborative writing project.

Boundary objects, according to Star (2010), are a sort of arrangement—institutional, conceptual or physical—that allow those different arrangements to be in contact, exchange and perhaps work together through the communities that take actions in and around them. However, the form they take is not arbitrary. They are emergent infrastructures that have arisen due to information needs and work requirements among groups who wish to cooperate. Star and Griesemer (1989) — who arguably coined the concept of boundary objects note that they have three features, which in our view, commons also have: (1) interpretive flexibility, (2) emergent structures of information and work process needs and arrangements (e.g., institutions, networks, markets), and (3) the dynamic between ill-structured and more tailored uses of the object (e.g., multi-functional). The third point means there are different uses that can serve multiple functions where they (e.g., commons) dynamically push and pull between two or more adjacent objects or realities without ever being a perfect

G. Caniglia
University of Helsinki, Helisinki, Finland
e-mail: guido.caniglia@helsinki.fi

F. Clement
Dynafor, INRAE, Toulouse University, Toulouse, France
e-mail: floriane.clement@inrae.fr

E. R. Devlin
University of the Highlands and Islands, Inverness, Scotland, UK

G. Epstein
Blue Range Labs and School of Environment, Resources and Sustainability, University of Waterloo, Waterloo, Canada
e-mail: graham.epstein@uwaterloo.ca

S. Jentoft
Norwegian College of Fishery Science, UiT - The Arctic University of Norway, Tromsø, Norway
e-mail: svein.jentoft@uit.no

C. Luederitz
Department of Management, University of Quebec at Trois-Rivières, Trois-Rivières, Canada
e-mail: Christopher.Luederitz@uqtr.ca

J. Pittman
School of Planning, Faculty of Environment, University of Waterloo, Waterloo, Canada
e-mail: jpittman@uwaterloo.ca

Table 1 Tensions or dichotomies that can arise in the commons, but within which they also act as boundary objects for constructive engagement through dialogue about ethical values, plurality, and co-existence

People	↔	Nature
Research	↔	Practice
Descriptive	↔	Normative
Freedom	↔	Responsibility
Local needs	↔	Global limits
Insider	↔	Outsider
Universal knowledge	↔	Place-based realities
Western/northern	↔	Non-Western/southern/ indigenous
Sufficiency	↔	Opportunity
Role of states	↔	Role of civil society
Public	↔	Private
Open	↔	Closed
Past	↔	Future
Short-term gain	↔	Long-term perspective

fit to either, enabling co-existence or translocation, or transformation. In these three features, ethical questions arise, including: who defines the commons? who is responsible for stewarding commons? how should the risks and benefits of the commons be governed?

If the commons are interpretatively flexible, what can or could they be? The commons literature struggles with clarity in relation to defining the plural types and interpretations of what the commons are—because the term 'commons' is generally representative of many specific shared objects, spaces, and institutions including physical objects, forms of governance rights and institutions, and ideological positions (Table 2). Traditional interpretations lean on commons as a type of good with certain kinds of inherent properties such as common-pool resources (subtractability; difficult to exclude use) or public goods (non-subtractable; difficult to exclude use), and these inherent features define their likely problems (i.e., overharvesting, and under provision/ free-riding respectively). There are many assumptions made in predicting or assuming that a type of good or resource will have a certain type of problem due to its inherent properties. The existence of a fish is not an inherent problem, neither is a tree or water molecule. The problem framing arises in our assumption about how we might value those natural elements or non-human entities, and thus might intend to take actions for interacting with or using them (e.g., caring, praying, symbolizing, extracting, transporting). Perceived intentions about human relations to nature or natural resources normatively shape the problem frame of a commons (e.g., fish are scarce; fish are sacred; fish is valuable food; we need to agree about fish; we need rights to fish). A community who does not eat fish may not have a problem frame in relation to fish as a common-pool resource; but perhaps a symbolic or religious one. In contrast, a community with a long history of local subsistence fishing may have valued traditions, food cultures, and practical experiences for mitigating overuse and incentivizing community governance strategies, and also argue strongly to keep those rights and autonomous self-governance of their fish. Intentions shape beliefs and guide actions and shape outcomes. An important ethical question to ask is: what are the ethically meaningful aspects of the actions? Is it the intentions, the way the actions are decided/enable, or the outcomes that matter most? Referring back to Table 3, and different ethical approaches, Kant (deontology) may have argued the intentions to act following moral duties is the most important, whereas Ubuntu, African Palaver, or contractualism perspectives may argue it is about how the action was decided in relation to others. Consequentialists might argue that only the outcome matters, as a measure of whether the action could be considered ethically good.

Other interpretations of the commons with different ethical implications take the form of assignable property rights regimes with shared ownership and co-governance

Table 2 Different forms of commons, their associated terminology and potential ethical questions

Forms of commons	Key terms or narratives	Questions in relation to ethics
Commons as a resistance ideology	Degrowth Anti-capitalist Anti-colonial Social movements/change	How can current power imbalances be shifted? What ethical values underpin the current state of affairs? What ethical values do we envision in the future? How can/do we bring these values about?
Commons as relational communities	Commoning Solidarity and care Connection to nature Sense of togetherness	How should we govern together? How can equity and justice be fostered? Can power structures be inclusive and transparent? How can decision-making be participatory? How are we dependent on and relational with nature?
Commons as property regimes	Governance Bundles of rights Institutions	What are the governance intentions and mechanisms? Are commons more ethical than alternatives? Who should have rights? What is the structure and design of governance? How does institutional development and change happen?
Commons as resources	Common-pool resources Public goods	How and why are commons problems framed? Who should benefit from the commons? Who should steward and be responsible for commons?

Table 3 Here we list simplified and interpreted claims of different ethical theories and broader cultural approaches to ethics originating from different time periods and world regions. When thinking about the commons in the context of each theory or approach, recognizable features and terminology can be observed as being associated with different aspects of the commons and commoning literature. The claims are a brief interpretation of each approach, and do not intend to represent the complexity and depth that each approach embodies. It is not our intention to misrepresent or misuse the ethical theories or cultural approaches in any way. In the spirit of building ethical awareness, we believe it is important to acknowledge the many different cultural practices and their contributions to ethical thinking beyond Western/Northern academic theorizing

Ethics approach	Theory or practice	Origin/ notable figures	Simplified ethical claims
Discourse ethics	Specific theory	Habermas Apel (Germany)	Actions are ethically good when … they result from a discourse following defined rules of equality, respect and fair participation of all persons concerned.
Virtue ethics	Specific theory	Confucius Plato Aristotle	Actions are ethically good when… they arise from virtues such as generosity, justice or compassion. Ethically good actions are those that a virtuous person does.
Consequential-ism	Specific theory	Bentham (UK) Mill (UK)	Actions are ethically good when… their consequences maximize Well-being and minimize harm for the greatest number.
Feminist ethics	Specific theory	1970s (USA/ Europe)	Actions are ethically good when… they address power imbalances, support relational care, reorganize reproductive labor, and promote inclusivity and equity.
Contractual- Ism	Specific theory	Hobbes (UK) Rawls (USA) Scanlon (USA)	Actions are ethically good when… they emerge from agreements that are justifiable to all affected parties, ensuring fairness, respect agency, and mutual benefit.
Deontology	Specific theory	Kant (Germany)	Actions are ethically good when… they are guided by maxims which can be universalized according to the categorical imperative. Moral duties apply to everyone such as respecting human dignity and autonomy.
Life ethics	Specific theory	Center for Life Ethics (Germany)	Actions are ethically good when… respecting or fostering the unfolding, plenitude, beauty, and embeddedness of all life.

(continued)

Table 3 (continued)

Ethics approach	Theory or practice	Origin/ notable figures	Simplified ethical claims
Environmental ethics	Specific theory	Carson (USA) Leopold (USA) Naess (Norway)	Actions are ethically good when…moral consideration extends beyond humans to nature and more-than-human entities to recognize the interdependencies and relational connections between the biological community.
Ubuntu ethics	Broad cultural practice	Indigenous peoples of southern Africa	Actions are ethically good when… they promote interconnectedness, mutual care, and the Well-being of all in the community, reflecting "I am because we are."
African palaver ethics	Broad cultural practice	Indigenous peoples of sub-Saharan Africa	Actions are ethically good when… they originate from a consensus that is reached by dialogue in spaces for open communication for all persons of the community.
Buddhist ethics	Broad cultural practice	~500 BCE (India)	Actions are ethically good when… they originate from infinite compassion that is free of desire and are aligned to purity and perfection of mind. This entails awareness of interdependence with all life.
Buen Vivir	Broad cultural practice	Andean indigenous peoples (South America)	Actions are ethically good when… actions pursue reciprocity and care, harmony with nature, collective wellbeing, sufficiency, plurality, and relationality.
Seven grandfather teachings	Broad cultural practice	Anishinaabe indigenous peoples (North America)	Actions are ethically good when… they are guided by the principles of love, respect, bravery, truth, honesty, humility, and wisdom.

responsibilities (Table 2). There are many documented variations of common property governance, often distinguished by their bundles of rights, which clarifies who has rights to access, withdraw, management, exclude (e.g., exclude others from rights) and alienate (i.e., sell or trade the rights) (Schlager & Ostrom, 1992). Different stakeholders may have different bundles or combinations of rights. More recently, commons have been defined being the community itself, which are (re)produced through shared rules and practices (de Angelis and Harvie, 2013, Federici, 2011, Fournier, 2013). Yet, others view commons as a political concept, which tends to embody two similar (and often interlinked) but contrasting narratives, as observed informally in the literature. The first is commons as a means of collective resistance to historical and current oppressive governance regimes, often characterized and manifested within anti-colonial, anti-racist, and anti-capitalist social movements, commoning and their associated ideological positions. It also goes beyond a defensive movement against commons enclosure to embody a creative political and economic alternative through the production of shared space, economies, and ways of governing. The second narrative is more relational in its discourse, also linked to commoning, but appealing to ideas of solidarity, care, community resilience, connection to nature, and a sense of togetherness. Depending on the interpretative flexibility engaged with, different ethical tensions and values will come into play, shaping the intentions of our research and actions.

Commons governance and commoning are perhaps most famously associated with observations of effective self-governance, and the diversity of institutional arrangements (i.e., set of rules, norms, networks, organizations) that emerge from when local problems need to be solved by local people across unique contexts. If there is one thing the commons literature may agree on, is that there is no one-size-fits-all set of governance approaches. Each rule, norm, resource distribution mechanism and leadership structure will imply certain ethical values in it, and with it perhaps different intentions behind them despite similar structural features. The idea that commons can be conceptually interpreted as boundary objects due to this feature is no surprise. Although this may also be a feature between many different concepts and ethics, commons provide a clear vantage point over many diverse challenges and conflicts that highlight the need for building ethical awareness.

2 Building Ethical Awareness

Ethics is necessarily applied in all intentional human actions, including commons governance and commoning, which generally refers to the social actions related to commons governance and the production communities. However, what is potentially ethically good about using and sustaining commons, or engaging in commoning practices, is not always clear. The ethics of the commons are often normatively implied in much of the literature, but not often linked to an ethical theory or philosophical foundation, or explicitly discussed as ethics. In the field, there is no well-established ethical framework or theory associated with the commons. Due to their collective and shared nature—and the dependence of global society on their health and sustainability - the commons offer an opportunity to further engage with ethical questions, and the relationships and practices they entail. We argue that the commons can be viewed as boundary objects with interpretive flexibility, emergent institutions, and shifting goals (Leigh Star, 2010), as entities and processes that can facilitate plural exchanges and discourse on ethics. We see commons as an opportunity space to address the tensions and conflicts that can arise between people who share vital spaces, resources, institutions and communities, but who likely have differing goals, values, capacities and practices. This can be at all levels from local to global. Sitting with and in these tensions is key for recognizing ethical plurality and building ethical awareness. Many scholars, project teams, and practitioners experience these tensions, but may not have a toolkit to constructively engage with them. The ability to navigate these values and to understand their nuances while fostering and engaging in dialogue about what may be right or wrong is essential for ethically aware science and practice. The need for building ethical awareness is not unique to the commons. But the shared and rival nature of many commons, and the desire to establish more common property governance—and the rising ambition to govern society through forms of commoning—makes the ethical dimensions of the commons more important to clarify.

Ethics are intertwined with the goals and means of commons (i.e., the why and how). They become reflected in our intentions and actions and often also in their consequences, but importantly reflect intentions. The same actions could be taken elsewhere with different intentions, and thus be driven by different ethical values. This is why it is important to understand the values underpinning the concepts and arguments used to justify commons governance and commoning. Different ethical theories and ethical practices—of which there are a diverse spectrum of Northern, Western, Eastern, Southern, and Indigenous approaches—with long histories of commons engagement can help build a toolkit for navigating issues in current scholarship and practice. However, these different approaches can only be useful if we build ethical awareness about them. Some of these approaches could include African Palaver ethics, the Seven Grandfather Teachings, Ubuntu ethics, Neo-Confucian ethics, Feminist ethics, Buddhist ethics, consequentialism, virtue ethics, deontology, or contractualism (see Table 3 for an exploratory overview of how ethical claims pertain to the commons). A useful way to build ethical awareness on the commons in relation to these approaches, out of each which some are - especially the Western one - conceptualized as ethical "theories", is to ask what each approach would value about the commons from their perspective.

Recognizing the plurality of ethical approaches that exist offers the basis to develop a toolkit to navigate the ethical implications of the commons and commoning. An ethical toolkit is the set of practical resources that an individual or groups have, including key concepts, principles, and methods to comprehend and analyze ethical issues, and to constructively navigate moral conflicts (Baggini & Fosl, 2014). Each commons and governance process will differ, and require a contextualized approach given the problems that arise and the people who are involved and their interests. Clarifying the ethical issues at hand, recognizing the ethical values behind the respective intentions, and choosing actions that are informed by ethical values after critical reflection, is important for how we do our research as well as how we evaluate governance of the commons and commoning (see Box 1 for exemplary reflective questions). We encourage scholars to build up ethical awareness surrounding the commons with more explicit and justifiable language, concepts, theories, and practices that can clarify their actions and foster dialogue. Importantly, a toolkit, on the other hand, does not suggest that the justification of our intentions and actions is arbitrary or not more than relative to a unique context. Putting decision-making and ethical values into dialogue—whether in research or practice - provides a space where plural perspectives are valued, can be justified and clarified. Nonetheless, we do not intend to simply instrumentalize the use of ethical theories and approaches, but rather to encourage exploration of the how the concepts and terminology used in the commons and commoning literature link back to philosophical foundations of different approaches. Also, how research can be designed to better capture, analyze, and communicate ethical aspects of empirical research and its possible consequences from the beginning.

Overall, building ethical awareness around the commons requires expanding the field of commons research into new inter- and transdisciplinary areas to facilitate shared learning. From an interdisciplinary perspective, philosophy and the commons can be brought closer together to strengthen research on the commons with existing ethical theories and debates. The field of commons research has long been recognized and led by interdisciplinary scholars (Poteete et al., 2010). However, collaborations have been mostly limited to scientists with roots in political science, economics, geography, and sociology (van Laerhoven et al., 2020). There is room to expand to philosophy, arts and humanities, history, religion, anthropology, future studies, environmental and natural sciences. Furthermore, many ethical approaches only exist as practices, and have not been formalized academically into writing, for example, in many Indigenous, religious or other non-Western approaches. Here, transdisciplinary approaches can help to better understand how practices in and around the commons are related to ethical values, when effectively engaged with and reflected on throughout the process (Partelow et al., 2024). Taking participatory methodological approaches such as those in sustainability science, co-production and participatory action research can be useful and should be seriously engaged with for integrative study designs (Bandola-Gill et al., 2023; Miller & Wyborn, 2020). Transdisciplinarity nonetheless raises many ethical research issues regarding community participation, engagement with power dynamics and intentions, data protection, knowledge ownership, and scientific integrity (Manlosa et al., 2023; Strand, 2024). Traditional science-led approaches also need to play a strong role in documenting and analyzing both the descriptive and normative dimensions of commons governance and commoning; also abiding by the necessary ethical research protocols. All research approaches can add value, perhaps most importantly when they are in constructive dialogue with each other.

Box 1. Reflective questions for building ethical awareness on commons governance and commoning

> What are the ethical or normative value assumptions I have about the commons?
> What are the ethical values driving my research?
> What are the ethical claims behind the concepts I use in my research?
> How can I be clearer about my assumptions, positionality and values in my research?
> How do my ethical values relate/compare to those of the people/communities I am working with?
> What power relations and dynamics exist between my values and those held by others?

How could my ethical values and assumptions negatively affect those whom I work with?
What am I doing to avoid unintentional negative outcomes through my research?

There is not singular ethical theory or approach for the commons, many have resonating elements and claims (i.e., from Table 3). Just as there are no governance panaceas (Brock & Carpenter, 2007; Ostrom, 2007), there are no universally applicable ethically approaches for the commons either. The same answer may apply to privatization or state-led approaches (Partelow et al., 2019). It depends, means what are the goals, outcomes, values, and processes of interest? Commons are not one thing. They are a set of social processes and institutions evolved in a particular place, with a certain relationality embedded into the human-nature dynamics of that place. Thus, the commons are not likely to be universal either. Nonetheless, to even begin exploring questions of ethical orientation, and what might be ethically preferred, there is a need to clarify the overarching governance mechanisms and intention of the commons. This is that commons create the conditions and responsibility for self-governance or co-governance, either partially or fully. How does this claim relate to the ethical approaches, for example, in Table 3? In practice, and in most places, where commons exist, it is partially given considering the role of states in recognizing and allocating rights and financing, as well as the role that adjacent or embedded markets and private rights play in shaping commons and common property governance. Yet, the conditions and responsibility for self-governance remain relevant for recognizing the in-group value of self-organization processes and how these relational human-nature processes evolve into cultural norms and place-adapted rules and decisions-making processes over time, which are generally viewed as ethically good actions in the commons literature. Subsequent to this question, are many other questions about how and why self-governance should or could be done normatively. Ethically relevant actions such as transparency, care, and participation are often positioned normatively as social equitable or just. Other key questions are: what commons is being governed and how is it defined? Who is responsible for self-governance of that commons? How does the process of self-governance emerge and operate? Who can receive or be impacted by risks, burdens and benefits? There are of course many other questions at a high level, and subsequent to those, of high relevance for the ethical values of the commons. For example, the question of whether certain governance mechanisms and practices can be applied or exported elsewhere, to other commons? How place-based and uniquely relational are different commons? Can we transfer or scale principles, while also recognizing that commons are situated, localized and plural? There are also commons that are global in scale by their nature, for example, breathable air, rainwater, the internet, and the climate, even though they are governed in smaller socially determined units where intentions and impacts can be localized. Understanding commonalities in the ethical underpinnings across different types of commons and commoning may shed further light on these practices. Nonetheless, taking a one-size-fits-all approach merely reduces the complexity and nuances that characterize diverse circumstances, places and people that give commons meaning and value. An important feature, in fact, is that commons are not separate from the communities that use and govern them. They are interwoven and co-evolved. Commons are often portrayed as socially constructed communities, but interwoven as human-nature relations. Being relationally interdependent means they can only exist when the other exists. Commons are defined by the fact that they are shared by people who form communities in relation to them. As a result, the ethics of the commons as a governance solution very much depends upon the interests, values and needs of communities—and capacities for relationality within the communities and commons—which are likely to be pluralistic across different members of those communities.

3 Research Integrity and Teaching Ethics on the Commons

Research ethics are a different topic than the ethical issues of the commons governance and commoning as objects and actions for analysis in civil society. A key question is: do research ethics on the commons need different considerations than on other topics? Research ethics include both formal mandates and emerging informal norms, and often span multiple ethical perspectives throughout the research process. Many universities have institutional review boards or ethics committees, where researchers are often required to submit their research plans to be assessed as a means of entering a dialogue about potential ethical issues, or to assess coherence with established ethical codes of conduct. While these institutional structures are an essential part of the integrity of universities and research organizations outside of universities, there are also many critiques including their focus on legal compliance over practical ethical issues, the lack of place-based cultural considerations, their limited ability to represent diverse affected groups, inconsistency in decision-making, and the need to only address ethical issues before and not during or after the research (Boden et al., 2009). Furthermore, it can be difficult for researches to comply with all regulations, even practically, given the many regulations at different regulatory levels, and even conflicting differences across countries. For example, different research sample handling protocols or co-authorship rules between collaborating countries (Rochmyaningsih, 2018). Due to these reasons, many research fields have made advancements to develop ethical guidelines and practices themselves as an alternative or in addition, which also have benefits and shortcomings.

Commons research is primarily led by social scientists who use social science methods, focusing heavily on people, communities, and the social dimensions of governance in relation to culture, economy, politics and place. It is clear that research ethics issues needing consideration will differ by the type of research approach taken. Three core approaches embody the field, each having their own research ethics considerations: (1) external or researcher-led approaches, (2) community-led, and (3) co-production, collaborative or transdisciplinary approaches. For each, being transparent and recognizing the positionality of the research in the process is a central starting point. Engaging with and reflecting on ethical issues and values throughout the research process is essential. Building ethical awareness about the intentions of the research, and how to design studies in a way that not only effectively answers research questions but that also aligns with the ethical values of the intentions (Box 2).

Box 2. Potential questions for research ethics

What are the intentions of our research at different levels (i.e., individually; institutionally)?
What research actions and approaches best reflect our intentions?
How can transparency enable dialogue about the benefits and challenges of different approaches?
How and why do we teach the ideas and concepts of the commons?
What are our intentions to communicate our science?
Who does our research affect and in what ways?

The ethics of how and why we teach the commons is another dimension of academia that is rarely discussed. It is worth mentioning here although it is not a topic that is addressed by a specific chapter in this book. Many of the same questions can be asked about teaching and communicating the commons. What are our intentions for teaching at different levels? What teaching actions align with our intentions? Is the outcome all that matters? Or that process of how we teach? If the two central goals of academics are to create and disseminate knowledge, teaching is one key form of dissemination, but not the only one. Many scientists also engage substantially in civil society in their scientific capacities. The role and involvement of scientists in social movements, policy consultation, political activism, government and local community initiatives is arguably increasing. Many scientists would like to use their knowledge, and power as knowledge holders, from within recognized institutions to directly advocate for changes in society. Many graduate and doctoral students get into science - or switch to focus on the social sciences - with the intention to have a greater impact in society, for example. These issues are by no means unique to the field of the commons, but are nonetheless a central feature that would benefit from building ethical awareness and having a toolkit that can foster constructive dialogue at different institutional levels about intentions, actions, and outcomes, and what ethical approaches are guiding these activities.

4 Reflection Points for Building Ethical Awareness

To conclude this final chapter, we offer reflective points as potential guidance on how to build ethical awareness. The points below are a reflection of our experience exploring ethical questions and becoming ethically curious in our research. This is by no means a comprehensive list, and we encourage you to discuss these issues with colleagues in the context of your research, organization, and interests, as a way to catalyze interest in the field. The specific chapters in this book have offered insights on specific thematic topics and perspectives. Here we extend an opportunity for broader reflections

- Engage with the reflective questions in this chapter in the context of your own research and teaching.
- Reflect on the ethical values supporting the intentions of your research and your organization.
- Where possible pursue transparency in the intentions and motivations of the questions you ask, methods you use and communication strategies you adopt.
- Be curious about ethical theories and concepts in philosophy, and how they can be linked to your research.
- Explore your own ethical values, both personally and professionally. Why do you have these values? What drives or inspires a change in those values?
- Many ethical perspectives and values exist—which help to clarify the ethics of different actions in different contexts. Embracing ethical plurality requires open-mindedness and curiosity for constructive engagement.
- Observe the actions of others (e.g., colleagues, research subjects, friends and family) when commoning/dealing with commons, and explore the ethical values that might be motivating them.
- How have commons become a boundary object in your research? Have commons (in any form) enabled you to explore tensions between conflicting or competing values, capacities or practices?
- How do you view the role of science in society? How do you view yourself as an individual within that system?

References

Bandola-Gill, J., Arthur, M., & Leng, R. I. (2023). What is co-production? Conceptualising and understanding co-production of knowledge and policy across different theoretical perspectives. *Evidence and Policy, 19*(2), 275–298. https://doi.org/10.1332/174426421X16420955772641

Boden, R., Epstein, D., & Latimer, J. (2009). Accounting for ethos or programmes for conduct? *The brave new world of research ethics committees.*

Brock, W. A., & Carpenter, S. R. (2007). Panaceas and diversification of environmental policy. *Proceedings of the National Academy of Sciences of the United States of America, 104*(39), 15206–15211. https://doi.org/10.1073/pnas.0702096104

van Laerhoven, F., Schoon, M., & Villamayor-Tomas, S. (2020). Celebrating the 30th anniversary of ostrom's governing the commons: Traditions and trends in the study of the commons, revisited. *International Journal of the Commons, 14*(1), 208–224. https://doi.org/10.5334/ijc.1030

Leigh Star, S. (2010). This is not a boundary object: Reflections on the origin of a concept. *Science, Technology & Human Values, 35*, 601–617. https://doi.org/10.1177/0162243910377624

Manlosa, A. O., Partelow, S., Jiren, T. S., & Riechers, M. (2023). The role of institutions in food system transformations: Lessons learned from transdisciplinary engagements in Ethiopia, The Philippines, and Indonesia. *Ecosystems and People, 19*(1). https://doi.org/10.1080/26395916.2022.2146753

Miller, C. A., & Wyborn, C. (2020). Co-production in global sustainability: Histories and theories. *Environmental Science and Policy, 113*, 88–95. https://doi.org/10.1016/j.envsci.2018.01.016

Ostrom, E. (2007). A diagnostic approach for going beyond panaceas. *Proceedings of the National Academy of Sciences of the United States of America, 104*(39), 15181–15187. https://doi.org/10.1073/pnas.0702288104

Partelow, S., Abson, D. J., Schlüter, A., Fernández-Giménez, M., Wehrden, H. von, & Collier, N. (2019). Privatizing the commons: New approaches need broader evaluative criteria for sustainability. International Journal of the Commons, 13(1), 706–747. https://doi.org/10.18352/ijc.777.

Partelow, S., Luederitz, C., Huang, Y.-S., von Wehrden, H., & Woopen, C. (2024). Building ethical awareness to strengthen co-production for transformation. *Sustainability Science.* https://doi.org/10.1007/s11625-024-01582-7

Poteete, A., Janssen, M., & a.,·& Ostrom, E. (2010). *WORKING TOGETHER: Collective action, the commons and multiple methods in practice.* Princeton Universty Press.

Rochmyaningsih, D. (2018). Indonesia plans strict foreign research laws. *Nature, 557*(7706), 476. https://doi.org/10.1038/d41586-018-05001-7

Strand, M. (2024). Advancing transdisciplinarity as an epistemology for more ethical engagements with fishing communities. *ICES Journal of Marine Science.* https://doi.org/10.1093/icesjms/fsae148

Stefan Partelow is a Professor at the Center for Life Ethics, University of Bonn. His research focuses on commons governance, collective action, and sustainability transformations in coastal and marine social–ecological systems. He combines empirical research and ethical reflection to promote just and adaptive approaches to environmental governance.

Guido Caniglia is an Associate Professor of Social Policy at the University of Helsinki, affiliated with HELSUS and the Helsinki Inequality Initiative. Holding dual PhDs in Philosophy and Biology, his interdisciplinary research spans sustainability science, epistemic justice, and transdisciplinary collaboration. He leads the ERC-funded WEIRD project, which examines sustainability through disability and queer perspectives.

Floriane Clément is a researcher at Dynafor, INRAE, Toulouse University. Her work examines agri-environmental governance and the role of farm advisors as interface bureaucrats in implementing agroecological policies. She applies feminist political ecology and institutional analysis to understand the everyday practices and tensions in agricultural transitions.

Erin Rizzato Devlin is affiliated with the University of the Highlands and Islands in Scotland, contributing to the Energy Innovation Team. Her research interests lie in marine social science and sustainability, particularly in rural and island contexts. She collaborates on interdisciplinary projects addressing energy transitions and community resilience.

Graham Epstein is a Research Assistant Professor at the University of Waterloo's School of Environment, Resources and Sustainability. His research explores the human dimensions of conservation and natural resource governance, focusing on collective action, compliance, and institutional performance. He collaborates with NGOs and government partners to inform policy design.

Svein Jentoft is a sociologist and Professor Emeritus whose work focuses on marine resource management; small-scale fisheries; and coastal communities, both in the Global North and South.

He is a founding member of the international network Too Big To Ignore (TBTI) and has led initiatives on governance of fisheries, particularly relating to SSF Guidelines (Voluntary Guidelines for Securing Sustainable Small-Scale Fisheries). His recent writing includes reflective and essay-style works about the human and community dimensions of small-scale fisheries, including Life Above Water and The Gift of Community.

Christopher Luederitz is a Professor in the Department of Management at the University of Quebec at Trois-Rivières. His work focuses on sustainability transformations in urban and organizational contexts, with particular attention to experimentation, living labs, and knowledge co-creation. He explores how universities, firms, and other institutions can foster systemic change for sustainability.

Jeremy Pittman is an associate professor in the School of Planning, University of Waterloo. His work focuses on governance in social–ecological systems, with emphasis on land–sea interfaces, species-at-risk conservation, and cross-scale institutional coordination. He uses mixed methods and network approaches to study how governance arrangements affect ecological and social outcomes.

Christiane Woopen is a Professor and the founding director of the Center for Life Ethics at the University of Bonn. She is a medical ethicist and philosopher whose work addresses health policy, reproductive and neuroethics, and the ethical implications of health technologies. Her research and leadership advance interdisciplinary approaches to life ethics and public responsibility.

Open Access This chapter is licensed under the terms of the Creative Commons Attribution-NonCommercial-NoDerivatives 4.0 International License (http://creativecommons.org/licenses/by-nc-nd/4.0/), which permits any noncommercial use, sharing, distribution and reproduction in any medium or format, as long as you give appropriate credit to the original author(s) and the source, provide a link to the Creative Commons license and indicate if you modified the licensed material. You do not have permission under this license to share adapted material derived from this chapter or parts of it.

The images or other third party material in this chapter are included in the chapter's Creative Commons license, unless indicated otherwise in a credit line to the material. If material is not included in the chapter's Creative Commons license and your intended use is not permitted by statutory regulation or exceeds the permitted use, you will need to obtain permission directly from the copyright holder.

[illegible]

[illegible]

[illegible]

[illegible]

[illegible]

[illegible]

MIX
Papier aus verantwortungsvollen Quellen
Paper from responsible sources
FSC® C105338

If you have any concerns about our products,
you can contact us on
ProductSafety@springernature.com

In case Publisher is established outside the EU,
the EU authorized representative is:
Springer Nature Customer Service Center GmbH
Europaplatz 3, 69115 Heidelberg, Germany

Printed by Libri Plureos GmbH
in Hamburg, Germany